USING the COMMON CRITERIA for IT SECURITY EVALUATION

OTHER AUERBACH PUBLICATIONS

The ABCs of IP Addressing
Gilbert Held
ISBN: 0-8493-1144-6

The ABCs of TCP/IP
Gilbert Held
ISBN: 0-8493-1463-1

Building an Information Security Awareness Program
Mark B. Desman
ISBN: 0-8493-0116-5

Building a Wireless Office
Gilbert Held
ISBN: 0-8493-1271-X

The Complete Book of Middleware
Judith Myerson
ISBN: 0-8493-1272-8

Computer Telephony Integration, 2nd Edition
William A. Yarberry, Jr.
ISBN: 0-8493-1438-0

Cyber Crime Investigator's Field Guide
Bruce Middleton
ISBN: 0-8493-1192-6

Cyber Forensics: A Field Manual for Collecting, Examining, and Preserving Evidence of Computer Crimes
Albert J. Marcella and Robert S. Greenfield, Editors
ISBN: 0-8493-0955-7

Global Information Warfare: How Businesses, Governments, and Others Achieve Objectives and Attain Competitive Advantages
Andy Jones, Gerald L. Kovacich, and Perry G. Luzwick
ISBN: 0-8493-1114-4

Information Security Architecture
Jan Killmeyer Tudor
ISBN: 0-8493-9988-2

Information Security Management Handbook, 4th Edition, Volume 1
Harold F. Tipton and Micki Krause, Editors
ISBN: 0-8493-9829-0

Information Security Management Handbook, 4th Edition, Volume 2
Harold F. Tipton and Micki Krause, Editors
ISBN: 0-8493-0800-3

Information Security Management Handbook, 4th Edition, Volume 3
Harold F. Tipton and Micki Krause, Editors
ISBN: 0-8493-1127-6

Information Security Management Handbook, 4th Edition, Volume 4
Harold F. Tipton and Micki Krause, Editors
ISBN: 0-8493-1518-2

Information Security Policies, Procedures, and Standards: Guidelines for Effective Information Security Management
Thomas R. Peltier
ISBN: 0-8493-1137-3

Information Security Risk Analysis
Thomas R. Peltier
ISBN: 0-8493-0880-1

A Practical Guide to Security Engineering and Information Assurance
Debra Herrmann
ISBN: 0-8493-1163-2

The Privacy Papers: Managing Technology and Consumers, Employee, and Legislative Action
Rebecca Herold
ISBN: 0-8493-1248-5

Secure Internet Practices: Best Practices for Securing Systems in the Internet and e-Business Age
Patrick McBride, Jody Patilla, Craig Robinson, Peter Thermos, and Edward P. Moser
ISBN: 0-8493-1239-6

Securing and Controlling Cisco Routers
Peter T. Davis
ISBN: 0-8493-1290-6

Securing E-Business Applications and Communications
Jonathan S. Held and John R. Bowers
ISBN: 0-8493-0963-8

Securing Windows NT/2000: From Policies to Firewalls
Michael A. Simonyi
ISBN: 0-8493-1261-2

Six Sigma Software Development
Christine B. Tayntor
ISBN: 0-8493-1193-4

A Technical Guide to IPSec Virtual Private Networks
James S. Tiller
ISBN: 0-8493-0876-3

Telecommunications Cost Management
Brian DiMarsico, Thomas Phelps IV, and William A. Yarberry, Jr.
ISBN: 0-8493-1101-2

USING the COMMON CRITERIA for IT SECURITY EVALUATION

DEBRA S. HERRMANN

A CRC Press Company
Boca Raton London New York Washington, D.C.

Library of Congress Cataloging-in-Publication Data

Herrmann, Debra S.
Using the Common Criteria for IT security evaluation / Debra S. Herrmann.
p. cm.
Includes bibliographical references and index.
ISBN 0-8493-1404-6 (alk. paper)
1. Telecommunication—Security measures—Standards. 2. Computer security—Standards. 3. Information technology—Standards. I. Title.

TK5102.85 .H47 2002
005.8—dc21 2002033250
CIP

Direct all inquiries to CRC Press LLC, 2000 N.W. Corporate Blvd., Boca Raton, Florida 33431.

International Standard Book Number 0-8493-1404-6
Library of Congress Card Number 2002033250
Printed in the United States of America 1 2 3 4 5 6 7 8 9 0
Printed on acid-free paper

Dedication

This book is dedicated to the victims of terrorist attacks
in Israel, New York City, Pennsylvania, and Washington, D.C.

Other Books by the Author

A Practical Guide to Security Engineering and Information Assurance (Auerbach Publications, 2001)

Software Safety and Reliability: Techniques, Approaches and Standards of Key Industrial Sectors (IEEE Computer Society Press, 1999)

Table of Contents

List of Exhibits

Chapter 2

Chapter 3

Chapter 4

Chapter 5

Chapter 1

Introduction

1.0 Background

In December 1999, ISO/IEC 15408, Parts 1–3 (Criteria for IT Security Evaluation), was approved as an international standard. The Common Criteria (CC) are considered *the* international standard for information technology (IT) security and provide a complete methodology, notation, and syntax for specifying security requirements, designing a security architecture, and verifying the security integrity of an "as built" product, system, or network. Roles and responsibilities for a variety of stakeholders are defined, such as:

- *Customers* — corporations, government agencies, and other organizations who want to acquire security products, systems, and networks
- *Developers* — (a) system integrators who implement or manage security systems and networks for customers, and (b) vendors who manufacture and sell commercial "off the shelf" (COTS) security products
- *Evaluators* — accredited Common Criteria Testing Laboratories, which perform an independent evaluation of the security integrity of a product, system, or network

Many organizations and government agencies require the use of CC-certified products and systems and use the CC methodology in their acquisition process. For example, in the United States, NSTISSP #11 (National Information Assurance Acquisition Policy)[75] mandated the use of CC-evaluated IT security products in critical infrastructure systems starting in July 2002.

Like ISO 9000, the Common Criteria have a mutual recognition agreement so that products certified in one country are recognized in another. As of June 2002, 15 countries have signed the mutual recognition agreement: Australia, Canada, Finland, France, Germany, Greece, Israel, Italy, the Netherlands, New Zealand, Norway, Spain, Sweden, the United Kingdom, and the United States.

1.1 Purpose

This book is a user's guide for the Criteria for IT Security Evaluation. It explains in detail how to understand, interpret, apply, and employ the Common Criteria methodology throughout the life of a system, including the acquisition and certification and accreditation (C&A) processes.

1.2 Scope

This book is limited to a discussion of ISO/IEC 15408, Parts 1–3 (Criteria for IT Security Evaluation) and how to use the Common Criteria within a generic system-development lifecycle and a generic procurement process. The terminology, concepts, techniques, activities, roles, and responsibilities comprising the Common Criteria methodology are emphasized.

1.3 Intended Audience

This book is written for program managers, product development managers, acquisition managers, security engineers, and system engineers responsible for the specification, design, development, integration, test and evaluation, or acquisition of IT security products and systems. A basic understanding of security engineering concepts and terminology is assumed; however, extensive security engineering experience is not expected.

The Common Criteria define three generic categories of stakeholders: customers, developers, and evaluators. In practice, these categories are further refined into customers or end users, IT product vendors, sponsors, Common Criteria Testing Laboratories (CCTLs), National Evaluation Authorities, and the Common Criteria Implementation Management Board (CCIMB). All six perspectives are captured in this book.

1.4 Organization

This book is organized into six chapters. Chapter 1 puts the book in context by explaining the purpose for which the book was written. Limitations on the scope of the subject matter of the book, the intended audience for whom the book was written, and the organization of the book are explained.

Chapter 2 introduces the Common Criteria (CC) by:

- Describing the historical events that led to their development
- Delineating the purpose and intended use of the CC and, conversely, situations not covered by the CC
- Explaining the major concepts and components of the CC methodology and how they work

- Illustrating how the CC relate to other well-known national and international standards
- Discussing the CC user community and stakeholders
- Looking at the future of the CC

Chapter 3 explains how to express security requirements through the instrument of a Protection Profile (PP) using the CC standardized methodology, syntax, and notation. The required content and format of a PP are discussed section by section. The perspective from which to read and interpret PPs is defined. In addition, the purpose, scope, and development of a PP are mapped to both a generic system lifecycle and a generic procurement process.

Chapter 4 explains how to design a security architecture, in response to a PP, through the instrument of a Security Target (ST) using the CC standardized methodology, syntax, and notation. The required content and format of an ST are discussed section by section. The perspective from which to read and interpret STs is defined. In addition, the purpose, scope, and development of an ST are mapped to both a generic system lifecycle and a generic procurement sequence.

Chapter 5 explains how to verify a security solution, whether a system or COTS product, using the CC/CEM (Common Evaluation Methodology). The conduct of security assurance activities is examined in detail, particularly why, how, when, and by whom these activities are conducted. Guidance is provided on how to interpret the results of security assurance activities. The relationship between these activities and a generic system lifecycle, as well as a generic procurement process, is explained. Finally, the role of security assurance activities during ongoing system operations and maintenance is highlighted.

Chapter 6 explores new and emerging concepts within the CC/CEM that are under discussion within the CC user community. These concepts have not yet been formally incorporated into the standard or methodology but are likely to be so in the near future.

Six informative annexes are also provided. Annex A is a glossary of acronyms and terms related to the Common Criteria. Annex B lists the sources that were consulted during the development of this book and provides pointers to other resources that may be of interest to the reader. Annex B is organized in three parts: (1) standards, regulations, and policy; (2) publications; and (3) online resources. Annex C cites the participants who have signed the Common Criteria Recognition Agreement (CCRA) and provides contact information for each country's National Evaluation Authority. Annex D lists organizations that are currently recognized as certified CCTLs in Australia and New Zealand, Canada, France, Germany, the United Kingdom, and the United States. Annex E lists organizations that are currently certified to operate Cryptographic Module Validation Program (CMVP) laboratories in Canada and the United States. Annex F is a glossary of CC three-character class and family mnemonics.

Chapter 2

What Are the Common Criteria?

This chapter introduces the Common Criteria (CC) by:

- Describing the historical events that led to their development
- Delineating the purpose and intended use of the CC and, conversely, situations not covered by the CC
- Explaining the major concepts and components of the CC methodology and how they work
- Illustrating how the CC relate to other well-known national and international standards
- Discussing the CC user community and stakeholders
- Looking at the future of the CC

2.0 History

The Common Criteria, referred to as "the standard for information security,"[117] represent the culmination of a 30-year saga involving multiple organizations from around the world. The major events are discussed below and summarized in Exhibit 1. A common misperception is that computer and network security began with the Internet. In fact, the need for and interest in computer security (or COMPUSEC) has been around as long as computers have. Primarily defense and intelligence systems employed COMPUSEC in the past. The intent was to prevent deliberate or inadvertent access to classified information by unauthorized personnel or the unauthorized manipulation of the computer and its associated peripheral devices that could lead to the compromise of classified information.[1,2] COMPUSEC principles were applied to the design, development, implementation, evaluation, operation, decommissioning, and sanitization of a system.

Exhibit 1. Time Line of Events Leading to the Development of the CC

Month/Year	Lead Organization	Standard/Project	Short Name
1/73	U.S. DoD	DoD 5200.28M, ADP Computer Security Manual — Techniques and Procedures for Implementing, Deactivating, Testing, and Evaluating Secure Resource Sharing ADP Systems	—
6/79	U.S. DoD	DoD 5200.28M, ADP Computer Security Manual — Techniques and Procedures for Implementing, Deactivating, Testing, and Evaluating Secure Resource Sharing ADP Systems, with 1st Amendment	—
8/83	U.S. DoD	CSC-STD-001-83, Trusted Computer System Evaluation Criteria, National Computer Security Center	TCSEC or *Orange Book*
12/85	U.S. DoD	DoD 5200.28-STD, Trusted Computer System Evaluation Criteria, National Computer Security Center	TCSEC or *Orange Book*
7/87	U.S. DoD	NCSC-TG-005, v1.0, Trusted Network Interpretation of the TCSEC, National Computer Security Center	TNI, part of Rainbow Series
8/90	U.S. DoD	NCSC-TG-011, v1.0, Trusted Network Interpretation of the TCSEC, National Computer Security Center	TNI, part of Rainbow Series
1990	ISO/IEC	JTC1 SC27 WG3 formed	—
3/91	U.K. CESG	UKSP01, U.K. IT Security Evaluation Scheme: Description of the Scheme, Communications-Electronics Security Group	—
4/91	U.S. DoD	NCSC-TG-021, v1.0, Trusted DBMS Interpretation of the TCSEC, National Computer Security Center	part of Rainbow Series
6/91	European Communities	Information Technology Security Evaluation Criteria (ITSEC), v1.2, Office for Official Publications of the European Communities	ITSEC
11/92	OECD	Guidelines for the Security of Information Systems, Organization for Economic Cooperation and Development	—
12/92	U.S. NIST and NSA	Federal Criteria for Information Technology Security, v1.0, Vols. I and II	Federal Criteria
1/93	Canadian CSE	The Canadian Trusted Computer Product Evaluation Criteria (CTCPEC), Canadian System Security Centre, Communications Security Establishment, v3.oe	CTCPEC
6/93	CC Sponsoring Organizations	CC Editing Board established	CCEB
12/93	ECMA	Secure Information Processing Versus the Concept of Product Evaluation, Technical Report ECMA TR/64, European Computer Manufacturers' Association	ECMA TR/64
1/96	CCEB	Committee draft 1.0 released	CC
1/96–10/97	—	Public review, trial evaluations	—
10/97	CCIMB	Committee draft 2.0 beta released	CC

Exhibit 1. Time Line of Events Leading to the Development of the CC (continued)

Month/Year	Lead Organization	Standard/Project	Short Name
11/97	CEMEB	CEM-97/017, Common Methodology for Information Technology Security Evaluation, Part 1: Introduction and General Model, v0.6	CEM Part 1
10/97–12/99	CCIMB with ISO/IEC JTC1 SC27 WG3	Formal comment resolution and balloting	CC
8/99	CEMEB	CEM-99/045, Common Methodology for Information Technology Security Evaluation, Part 2: Evaluation Methodology, v1.0	CEM Part 2
12/99	ISO/IEC	ISO/IEC 15408, Information technology — Security Techniques — Evaluation Criteria for IT Security, Parts 1–3 released	CC Parts 1–3
12/99 forward	CCIMB	Respond to Requests for Interpretations, issue final interpretations, incorporate final interpretations	—
5/00	Multiple	Common Criteria Recognition Agreement signed	CCRA
8/01	CEMEB	CEM-2001/0015, Common Methodology for Information Technology Security Evaluation, Part 2: Evaluation Methodology, Supplement: ALC_FLR — Flaw Remediation, v1.0	CEM Part 2 supplement

The *Orange Book* is often cited as the progenitor of the CC; actually the foundation for the CC was laid a decade earlier. One of the first COMPUSEC standards, DoD 5200.28-M (Techniques and Procedures for Implementing, Deactivating, Testing, and Evaluating Secure Resource-Sharing ADP Systems),[1] was issued in January 1973. An amended version was issued in June 1979.[2] DoD 5200.28-M defined the purpose of security testing and evaluation as:[1]

1. Develop and acquire methodologies, techniques, and standards for the analysis, testing, and evaluation of the security features of ADP systems.
2. Assist in the analysis, testing, and evaluation of the security features of ADP systems by developing factors for the Designated Approval Authority concerning the effectiveness of measures used to secure the ADP system in accordance with Section VI of DoD Directive 5200.28 and the provisions of the Manual.
3. Minimize duplication and overlapping effort, improve the effectiveness and economy of security operations, and provide for the approval and joint use of security testing and evaluation tools and equipment.

As shown in Section 2.2, these goals are quite similar to those of the Common Criteria.

The DoD 5200.28-M standard stated that the security testing and evaluation procedures "will be published following additional testing and coordination."[1] The result was the publication in 1983 of CSC-STD-001-83, the Trusted Computer System Evaluation Criteria (TCSEC),[3] commonly known as the *Orange Book*. A second version of this standard was issued in 1985.[4]

- CEM-99/045, Common Methodology for Information Technology Security Evaluation, Part 2: Evaluation Methodology, v1.0, August 1999
- CEM-2001/0015, Common Methodology for Information Technology Security Evaluation, Part 2: Evaluation Methodology, Supplement: ALC_FLR — Flaw Remediation, v1.1, February 2002

As the CEM becomes more mature, it too will become an ISO/IEC standard.

2.1 Purpose and Intended Use

The goal of the CC project was to develop a standardized methodology for specifying, designing, and evaluating IT products that perform security functions which would be widely recognized and yield consistent, repeatable results. In other words, the goal was to develop a full-lifecycle, consensus-based security engineering standard. Once this was achieved, it was thought, organizations could turn to commercial vendors for their security needs rather than having to rely solely on custom products which had lengthy development and evaluation cycles with unpredictable results. The quantity, quality, and cost effectiveness of commercially available IT security products would increase and the time to evaluate them would decrease, especially given the emergence of the global economy. As the CC *User Guide*[96] states:

> Adoption of the CC as a world standard and wide recognition of evaluation results will provide benefits to all parties:
>
> 1) a wider choice of evaluated products for consumers,
>
> 2) greater understanding of consumer requirements by developers, and
>
> 3) greater access to markets for developers.

There has been some confusion that the term "IT product" refers only to plug-and-play COTS products. In fact, the CC interpret the term "IT product" quite broadly:[110]

> ...a package of IT hardware, software, and/or firmware which provides functionality designed for use or incorporation within a multiplicity of systems. An IT product can be a single product or multiple IT products configured as an IT system, network, or solution to meet specific customer needs.

The standard gives several examples of IT products, such as operating systems, networks, distributed systems, and software applications.

The standard lists several items that are not covered and considered out of scope:[19]

- Administrative security measures and procedural controls
- Physical security
- Personnel security
- Use of evaluation results within a wider system assessment, such as certification and accreditation (C&A)
- Qualities of specific cryptographic algorithms

Administrative security measures and procedural controls generally associated with operational security (OPSEC) are not addressed by the CC/CEM. Likewise, the CC/CEM do not define how risk assessments should be conducted, even though the results of a risk assessment are required as an input to a Protection Profile (PP).[22] Physical security is addressed in a very limited context, that of restrictions on unauthorized physical access to security equipment and prevention of and resistance to unauthorized physical modification or substitution of such equipment.[20] (See functional security family FPT_PHS.) Personnel security issues are not covered at all; instead, they are generally handled by assumptions made in the PP. The CC/CEM do not address C&A processes or criteria. Doing so was specifically left to each country or government agency; however, it is expected that CC/CEM evaluation results will be used as input to C&A. The robustness of cryptographic algorithms or even which algorithms are acceptable is not discussed in the CC/CEM. Rather, the CC/CEM are limited to defining requirements for key management and cryptographic operations. (See functional security families FCS_CKM and FCS_COP.) Many issues not handled by the CC/CEM are covered by other national and international standards (see Section 2.3).

Four additional topics are not addressed by the CC/CEM or other national or international standards. First, system integration issues are not discussed, including the role of a system integration contractor, the integration of evaluated and non-evaluated products, and the integration of separately evaluated targets of evaluation (TOEs) (unless they are part of a composite TOE).

Second, CC evaluations take place in a laboratory, not the operational environment. Most large systems today are designed and implemented by system houses who integrate a variety of commercial and custom products or subsystems (COTS, GOTS, legacy systems, etc.) developed by multiple third parties. The integration of (1) security products with non-security products and (2) security products into an enterprise wide security architecture to provide the level of protection needed (and specified) is a major security challenge; that is, do the products work together accurately, effectively, and consistently? Many safety, reliability, and security problems are usually discovered during system integration and testing in the actual operational environment. If the CC is truly to become the "world standard and preferred method for security specifications and evaluations,"[117] the role of system integrators must be defined and guidance for conducting evaluations in the operational environment must be developed.

Third, the role of service organizations is not addressed, even though an assurance maintenance lifecycle is defined (see Class AMA, below, and Chapter 5). The two types of services organizations are:

1. Organizations that provide a "turn-key" system for consumers, with consumers being involved in specifying requirements but not design, development, operation, or maintenance
2. Organizations that perform the operation and preventive, adaptive, and corrective maintenance of a system

Most systems spend 20 percent of their life span in design and development and 80 percent of their life span in operation and maintenance. Except for "turn-key" systems, it is very rare for the same organization to perform both. Usually, one organization does the design and development of a system and another the operation and maintenance. Latent vulnerabilities and ineffective countermeasures will be exposed during the operation of a system. Consequently, the role of service organizations, whether they provide "turn-key" systems or perform operations and maintenance functions, must be defined in the CC/CEM. As Abrams[92] notes:

> The real world is populated with systems and services. Extending the CC to services is important if its utility is to be maximized.

Finally, publication of the Smart Card Security User's Group Protection Profile was a precedent-setting event in that this marked the first major application of the CC/CEM to chip technology. During this process, two limitations of the CC/CEM were discovered and reported:[100] (1) the need for CC components to deal with security application program interfaces (APIs), and (2) the fact that two CC components, FTP.ITC.1 and FPT_RVM.1, allow the initiation of a service to be defined but not its termination.

As the CC/CEM matures, these shortcomings and limitations will be overcome by the CC/CEM or a new related standard.

2.2 Major Components of the Methodology and How They Work

The three-part CC standard, ISO/IEC 15408, and the CEM are the two major components of the CC methodology, as shown in Exhibit 3.

2.2.1 The CC

Part 1 of ISO/IEC 15408 provides a brief history of the development of the CC and identifies the CC sponsoring organizations. Basic concepts and terminology are introduced. The CC methodology and how it corresponds to a generic system development lifecycle are described. This information forms

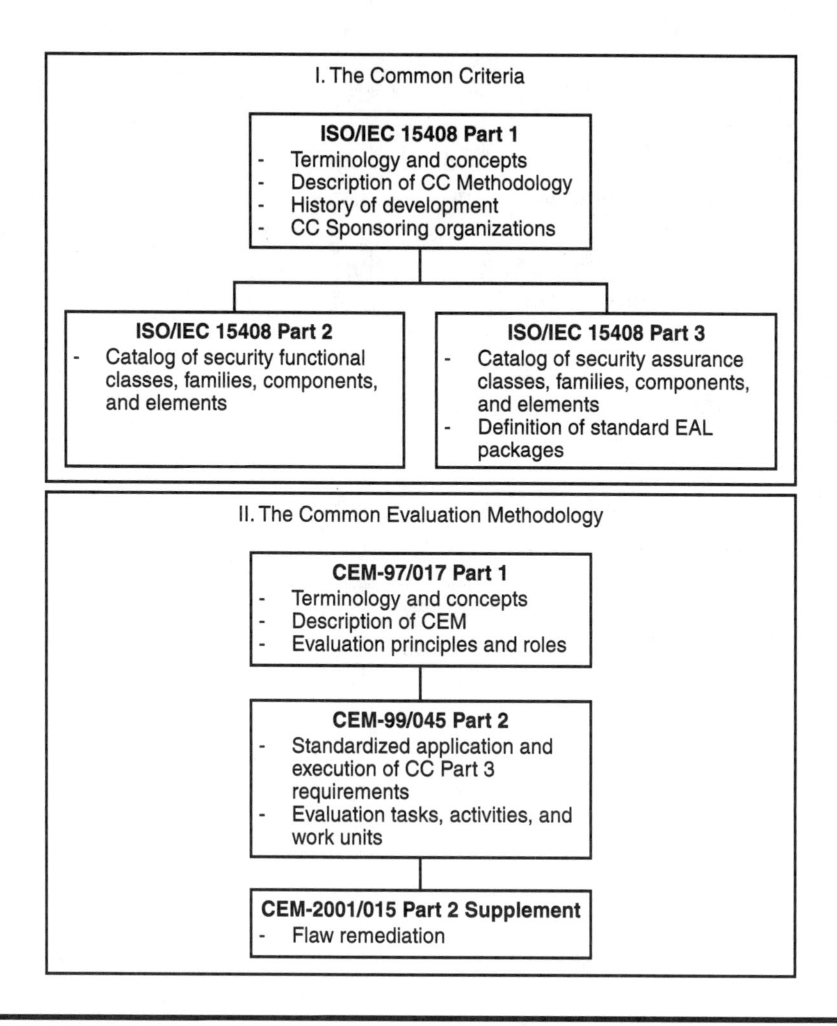

Exhibit 3. Major Components of the CC/CEM

the foundation necessary for understanding and applying Parts 2 and 3 of the standard.

Four key concepts are presented in Part 1 of the standard:

1. Protection Profiles (PPs)
2. Security Targets (STs)
3. Targets of evaluation (TOEs)
4. Packages

A Protection Profile (PP) is a formal document that expresses an *implementation-independent* set of security requirements, both functional and assurance, for an IT product that meets specific consumer needs.[19,23,110] The process of developing a PP helps consumers to elucidate, define, and validate their security requirements, the end result of which is used to: (1) communicate these requirements to potential developers, and (2) provide a foundation from which a Security Target can be developed and an evaluation conducted. Protection Profiles and their development are discussed in Chapter 3.

A Security Target (ST) is an *implementation-dependent* response to a PP that is used as the basis for developing a TOE. In other words, the PP specifies security functional and assurance requirements, while an ST provides a design that incorporates security mechanisms, features, and functions to fulfill these requirements. Security Targets and their development are discussed in Chapter 4.

A target of evaluation (TOE) is an IT product, system, or network and its associated administrator and user guidance documentation that is the subject of an evaluation.[19,23,24,110] A TOE is the physical implementation of an ST. The three types of TOEs are monolithic, component, and composite. A monolithic TOE is self-contained; it has no higher or lower divisions. A component TOE is the lowest level TOE in an IT product or system; it forms part of a composite TOE. In contrast, a composite TOE is the highest level TOE in an IT product or system; it is composed of multiple component TOEs.

A package is a set of components that are combined together to satisfy a subset of identified security objectives.[19] Packages are used to build PPs and STs. Packages can be a collection of functional or assurance requirements. Because they are a collection of low-level requirements or a subset of the total requirements for an IT product or system, packages are intended to be reusable. Evaluation assurance levels (EALs), discussed below and in Chapter 5, are examples of predefined packages.

As noted above, a PP represents a unique set of security functional and assurance requirements. Because these requirements are expressed in an implementation-independent manner, more than one implementation-dependent ST may be developed in response to a single PP. In other words, a one-to-many relationship exists between PPs and STs. Consumers have to determine which ST best meets their needs. A PP developed by one consumer may be reused by other consumers if they have identical requirements. TOE boundaries are defined in a PP; as a result, a PP may be written for a monolithic, component, or composite TOE. A one-to-one correspondence exists between an ST and a TOE, as a TOE is the physical implementation of a particular ST (see Exhibit 4).

Part 2 of ISO/IEC 15408 is a catalog of standardized security functional requirements (SFRs), which serve many purposes:[19,20,22] (1) describe the security behavior expected of a TOE, (2) meet the security objectives stated in a PP or ST, (3) specify security properties that users can detect by direct interaction with the TOE or by the response of the TOE to stimulus, (4) counter threats in the intended operational environment of the TOE, and (5) cover any identified organizational security policies and assumptions.

The CC organizes SFRs in a hierarchical structure of security functional:

- Classes
- Families
- Components
- Elements

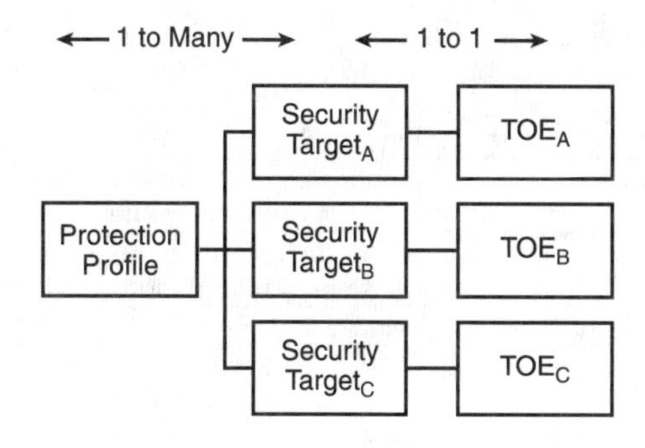

Exhibit 4. Relationship between PPs, STs, and TOEs

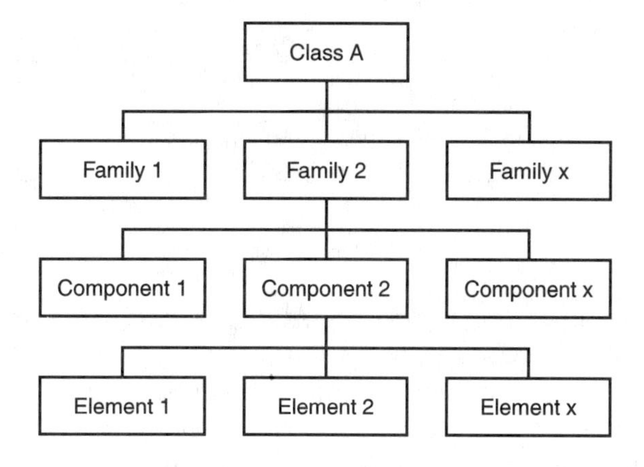

Exhibit 5. Relationship between Classes, Families, Components, and Elements

Part 2 defines 11 security functional classes, 67 security functional families, 138 security functional components, and 250 security functional elements. Exhibit 5 illustrates the relationship between classes, families, components, and elements.

A class is a grouping of security requirements that share a common focus; members of a class are referred to as families.[19] Each functional class is assigned a long name and a short three-character mnemonic beginning with an "F". The purpose of the functional class is described and a structure diagram is provided that depicts the family members. Exhibit 6 lists the security functional classes.

ISO/IEC 15408-2 defines 11 security functional classes. These classes are lateral to one another, with no hierarchical relationship among them. Accordingly, the standard presents the classes in alphabetical order. Classes represent the broadest spectrum of potential security functions that a consumer may need in an IT product. Classes are the highest level entity from which a consumer begins to select security functional requirements. It is not expected that a single IT product will contain SFRs from all classes.

Exhibit 6. Functional Security Classes

Short Name	Long Name	Purpose[20]
FAU	Security audit	monitor, capture, store, analyze, and report information related to security events
FCO	Communication	assure the identity of originators and recipients of transmitted information; nonrepudiation
FCS	Cryptographic support	manage and control operational use of cryptographic keys
FDP	User data protection	protect (1) user data, and the associated security attributes, within a TOE and (2) data that is imported, exported, and stored
FIA	Identification and authentication	ensure unambiguous identification of authorized users and the correct association of security attributes with users and subjects
FMT	Security management	manage security attributes, data, and functions and define security roles
FPR	Privacy	protect users against discovery and misuse of their identity
FPT	Protection of the TSF	maintain the integrity of the TSF management functions and data
FRU	Resource utilization	ensure availability of system resources through fault tolerance and the allocation of services by priority
FTA	TOE access	control user session establishment
FTP	Trusted path/channels	provide a trusted communication path between users and the TSF and between the TSF and other trusted IT products

The first class discussed is Security Audit, or FAU. FAU security functions are both proactive and reactive. Some FAU families focus on deterring security incidents by detecting actual, imminent, or potential security violations. Other FAU families support the traditional logging, storing, reporting, and analysis of audit trail data after the fact. The variety of FAU families and components accommodate the full range of audit needs for small, stand-alone systems or devices to large, complex distributed systems. In addition, provisions have been made to preempt the nemesis of security audit functions — misusing privileges or bypassing the audit function to prevent the capture of audit events.

The second class discussed is Communication, or FCO. As the name implies, FCO security functions pertain to the transportation of information. In particular, FCO families focus on the generation of evidence that transported information originated from a specific user or process and was indeed received by the designated recipient. These functions are referred to, respectively, as non-repudiation of origin and non-repudiation of receipt. The purpose of FCO functions is to ensure that information transported by a TOE is sent from and received by known subjects who cannot disavow participating in such communication afterwards.

Cryptographic Support, or FCS, is the third class discussed. This class applies to hardware, software, and firmware encryption. As noted in Section 2.1, the CC does not state which encryption algorithms or key lengths are acceptable. Instead, the FCS class focuses on cryptographic operation and key management — in other words, the secure use of encryption by a TOE. FCS components and elements are invoked whenever there is a need to generate or verify digital signatures, encrypt or decrypt data, perform secure functions, and so forth. Likewise, FCS components and elements are invoked to specify full-lifecycle key management activities, such as key generation, distribution, storage, recovery, and destruction.

User Data Protection, or FDP, defines a variety of requirements that protect user data. These requirements fall into four categories: (1) security policies that protect user data, such as access control and information flow control policies; (2) security functions that protect user data confidentiality and integrity during different types of online transactions, such as rollback, internal TOE transfers, and residual information protection after user data has been deleted or erased; (3) security functions that protect user data during offline transactions, such as import, export, and storage; and (4) security functions that protect user data confidentiality and integrity during inter-TOE security function communication.

The next class is Identification and Authentication, or FIA. As expected, this class defines requirements for performing and managing user identification and authentication functions. The correct identity of would-be users, whether authorized or unauthorized, is ascertained. The claimed identity of users is verified. Security attributes are correctly associated with each authorized user. Access control rights and privileges of each authorized user, relative to the TOE, are determined. In addition, the action to be taken following a specified number of authentication failures is defined.

The Security Management, or FMT, class specifies requirements for managing TOE security functions and their attributes and data. Conditions are defined for the establishment, revocation, and expiration of *bona fide* security attributes. Rules are established for when TOE security functions should or should not be invoked and by whom. Roles are created to separate the duties and responsibilities of security management personnel and the duties and responsibilities of security management personnel from other operational staff, such as network management.

Privacy requirements are specified through the FPR class. The purpose of these requirements is to protect users from having their identities discovered, misused, or associated with the use of TOE resources.

The FPT class contains requirements for protecting TOE security functions and TOE security function data. The underlying hardware and operating system upon which a TOE depends must execute as expected for TOE security functions to operate correctly. Consequently, requirements are defined for verifying this correct operation, such as self-tests, and detecting and responding to physical attacks upon the TOE. FPT defines requirements for the confidentiality, integrity, and availability of data exported by TOE security functions as well as data transferred or replicated internally. Conditions for trusted start-up and recovery of TOE security functions are stated. Mechanisms for detecting and preempting replayed messages, generating reliable time stamps, and synchronizing the timing of critical security functions are specified. Additional requirements ensure that TOE security policies are always invoked and enforced.

Resource utilization requirements, contained in the FRU class, ensure the availability of TOE resources for security functions. Fault tolerance requirements ensure that stated TOE capabilities will continue to operate correctly even when experiencing specified failure conditions. FRU allows TOE security functions to be assigned resource utilization priorities relative to other low-priority functions. Furthermore, FRU requirements allow resource utilization to be allocated among known users and subjects, thereby preventing resource monopolization and denial of service.

Target of evaluation access requirements are contained in the FTA class. Six types of requirements are developed that control different aspects of establishing a user session: (1) limiting the scope of user security attributes for a given session; (2) limiting multiple concurrent sessions by a single user; (3) locking and unlocking user sessions in response to given parameters; (4) displaying advisory banners about the use of TOE resources; (5) displaying a user's TOE access history, including successful and unsuccessful attempts; and (6) denying session establishment.

The FTP class develops requirements for trusted paths and trusted channels. These concepts are carried forward almost directly from the *Orange Book*. Requirements for trusted channels are invoked whenever there is a need to establish and maintain a secure communications channel between TOE security functions and other trusted IT products. In this instance, user or TOE security function data may need to be exchanged in order to perform security-critical functions, hence the need for a trusted channel. In contrast, requirements for a trusted path are invoked whenever there is a need to establish and maintain secure communications between users and TOE security functions. For both a trusted channel and a trusted path, the endpoints are identified and data is protected from unauthorized modification and disclosure while in transit. Communication can be initiated from either end of the trusted channel or trusted path.

A functional family is a grouping of SFRs that share security objectives but may differ in emphasis or rigor; the members of a family are referred to as components.[19] Each functional family is assigned a long name and a three-character mnemonic that is appended to the functional class mnemonic. Family behavior is described. Hierarchical relationships or ordering, if any, between members of a family are explained. Suggestions are made about potential OPSEC management activities and security events that are candidates to be audited. Exhibits 7 through 17 list functional security families by class.

Components are a specific set of security requirements that are constructed from elements; they are the smallest selectable set of elements that can be included in a PP, ST, or a package.[19] Components are assigned a long name and described. Hierarchical relationships between one component and another are identified. The short names for components consist of the class mnemonic, the family mnemonic, and a unique number.

An element is an indivisible security requirement that can be verified by an evaluation and the lowest level security requirement from which components are constructed.[19] One or more elements are stated verbatim for each component. Each element has a unique number that is appended to the component identifier. If a component has more than one element, all of them

Exhibit 7. FAU Functional Class: Security Audit

Family	Name	Function[20]	Component(s)	Name
FAU_ARP	Security audit automatic response	define action to be taken in response to a potential security violation	FAU_ARP.1	Security alarms
FAU_GEN	Security audit data generation	define which security events to record	FAU_GEN.1	Audit data generation
			FAU_GEN.2	User identity association
FAU_SAA	Security audit analysis	define requirements for automated analysis of security events	FAU_SAA.1	Potential violation analysis
			FAU_SAA.2	Profile based anomaly detection
			FAU_SAA.3	Simple attack heuristics
			FAU_SAA.4	Complex attack heuristics
FAU_SAR	Security audit review	define requirements for audit review tools	FAU_SAR.1	Audit review
			FAU_SAR.2	Restricted audit review
			FAU_SAR.3	Selectable audit review
FAU_SEL	Security audit event selection	define capability to select which events a TOE audits	FAU_SEL.1	Selective audit
FAU_STG	Security audit event storage	define capability to create and maintain a secure audit trail	FAU_STG.1	Protected audit trail storage
			FAU_STG.2	Guarantees of audit data availability
			FAU_STG.3	Action in case of possible audit data loss
			FAU_STG.4	Prevention of audit data loss

Exhibit 8. FCO Functional Class: Communication

Family	Name	Function[20]	Component(s)	Name
FCO_NRO	Non-repudiation of origin	Generate evidence for non-repudiation of origin.	FCO_NRO.1	Selective proof of origin
			FCO_NRO.2	Enforced proof of origin
FCO_NRR	Non-repudiation of receipt	Generate evidence for non-repudiation of receipt.	FCO_NRR.1	Selective proof of receipt
			FCO_NRR.2	Selective proof of receipt

Exhibit 9. FCS Functional Class: Cryptographic Support

Family	Name	Function[20]	Component(s)	Name
FCS_CKM	Cryptographic key management	specify full lifecycle key management activities	FCS_CKM.1	Cryptographic key generation
			FCS_CKM.2	Cryptographic key distribution
			FCS_CKM.3	Cryptographic key access
			FCS_CKM.4	Cryptographic key destruction
FCS_COP	Cryptographic operation	require cryptographic operations to be performed according to a specified algorithm and key size	FCS_COP.1	Cryptographic operation

must be used. Dependencies between elements are listed. Elements are the building blocks from which functional security requirements are specified in a PP. Exhibit 18 illustrates the standard CC notation for security functional classes, families, components and elements. Annex F provides a glossary of functional classes and families.

Part 3 of ISO/IEC 15408 is a catalog of standardized security assurance requirements, or SARs. SARs define the criteria for evaluating PPs, STs, and TOEs and the security assurance responsibilities and activities of developers and evaluators. The CC organize SARs in a hierarchical structure of security assurance classes, families, components, and elements. Part 3 defines 10 security assurance classes, 42 security assurance families, and 93 security assurance components.

A class is a grouping of security requirements that share a common focus; members of a class are referred to as families.[19] Each assurance class is assigned a long name and a short three-character mnemonic beginning with an "A". The purpose of the assurance class is described and a structure diagram is provided that depicts the family members. The three types of assurance classes are (1) those that are used for PP or ST validation, (2) those that are used for TOE conformance evaluation, and (3) those that are used to maintain security assurance after certification. Exhibit 19 lists the security assurance classes in alphabetical order and indicates their type.

ISO/IEC 15408-3 defines ten security assurance classes. Two classes, APE and ASE, evaluate PPs and STs, respectively. Seven classes verify that a TOE conforms to its PP and ST. One class, AMA, verifies that security assurance is being maintained between certification cycles. These classes are lateral to one another, with no hierarchical relationship among them. Accordingly, the standard presents the classes in alphabetical order. Classes represent the broadest spectrum of potential security assurance measures that a consumer may need to verify the integrity of the security functions performed by an IT product. Classes are the highest level entity from which a consumer begins to select security assurance requirements.

Exhibit 10. FDP Functional Class: User Data Protection

Family	Name	Function20	Component(s)	Name
FDP_ACC	Access control policy	define access control policies and the scope of control of each	FDP_ACC.1	Subset access control
			FDP_ACC.2	Complete access control
FDP_ACF	Access control functions	specify the implementation of each access control policy defined by FDP_ACC	FDP_ACF.1	Security attribute-based access control
FDP_DAU	Data authentication	provide guarantee of data validity	FDP_DAU.1	Basic data authentication
			FDP_DAU.2	Data authentication with identity of guarantor
FDP_ETC	Export to outside TSF control	specify limits on exporting user data and associating security attributes with exported user data	FDP_ETC.1	Export of user data without security attributes
			FDP_ETC.2	Export of user data with security attributes
FDP_IFC	Information flow control policy	define information flow policies and the scope of control of each	FDP_IFC.1	Subset information flow control
			FDP_IFC.2	Complete information flow control
FDP_IFF	Information flow control functions	specify rules for functions that implement information flow control	FDP_IFF.1	Simple security attributes
			FDP_IFF.2	Hierarchical security attributes
			FDP_IFF.3	Limited illicit information flows
			FDP_IFF.4	Partial elimination of illicit information flows
			FDP_IFF.5	No illicit information flows
			FDP_IFF.6	Illicit information flow monitoring
FDP_ITC	Import from outside TSF control	specify limits on importing user data and associating security attributes with imported user data	FDP_ITC.1	Import of user data without security attributes
			FDP_ITC.2	Import of user data with security attributes
FDP_ITT	Internal TOE transfer	specify requirements for protecting user data when it is transferred within a TOE	FDP_ITT.1	Basic internal transfer protection

Exhibit 10. FDP Functional Class: User Data Protection (continued)

Family	Name	Function[20]	Component(s)	Name
			FDP_ITT.2	Transmission separation by attribute
			FDP_ITT.3	Integrity monitoring
			FDP_ITT.4	Attribute based integrity monitoring
FDP_RIP	Residual information protection	ensure that deleted information is no longer accessible	FDP_RIP.1	Subset residual information protection
			FDP_RIP.2	Full residual information protection
FDP_ROL	Rollback	undo previous operation(s) in order to return to a known secure state	FDP_ROL.1	Basic rollback
			FDP_ROL.2	Advanced rollback
FDP_SDI	Stored data integrity	protect user data while it is stored within the TSC	FDP_SDI.1	Stored data integrity monitoring
			FDP_SDI.2	Stored data integrity monitoring and action
FDP_UCT	Inter-TSF user data confidentiality transfer protection	ensure confidentiality of user data while it is transferred between TOEs or users on different TOEs	FDP_UCT.1	Basic data exchange confidentiality
FDP_UIT	Inter-TSF user data integrity transfer protection	provide integrity for user data while in transit between the TSF and another trusted IT product	FDP_UIT.1	Data exchange integrity
			FDP_UIT.2	Source data exchange recovery
			FDP_UIT.3	Destination data exchange recovery

Protection Profile Evaluation, or APE, is the first security assurance class discussed. This class is invoked after a PP has been developed to determine whether a PP is adequate, complete, correct, and consistent. The activities defined in APE result in a formal evaluation of a PP. If the evaluation is successful, the PP is certified and becomes part of a National Evaluation Authority's PP registry. Evaluation goals are established for the first five sections of a PP:

- Is the PP identification information an accurate reflection of the PP?
- Is the TOE description coherent, internally consistent, and consistent with the remainder of the PP?
- Is the security environment in which the TOE will operate understood?

Exhibit 11. FIA Functional Class: Identification and Authentication

Family	Name	Function[20]	Component(s)	Name
FIA_AFL	Authentication failures	define the maximum number of unsuccessful authentication failures and the action to be taken when this number is reached	FIA_AFL.1	Authentication failure handling
FIA_ATD	User attribute definition	define security attributes that are associated with users	FIA_ATD.1	User attribute definition
FIA_SOS	Specification of secrets	enforce quality metrics on generated and provided secrets	FIA_SOS.1	Verification of secrets
			FIA_SOS.2	Generation of secrets
FIA_UAU	User authentication	define the types of user authentication mechanisms to be supported	FIA_UAU.1	Timing of authentication
			FIA_UAU.2	User authentication before any action
			FIA_UAU.3	Unforgeable authentication
			FIA_UAU.4	Single-use authentication mechanisms
			FIA_UAU.5	Multiple authentication mechanisms
			FIA_UAU.6	Reauthenticating
			FIA_UAU.7	Protected authentication feedback
FIA_UID	User identification	define conditions under which users have to be authenticated	FIA_UID.1	Timing of identification
			FIA_UID.2	User identification before any action
FIA_USB	User–subject binding	define requirements for associating a user's security attributes with a subject	FIA_USB.1	User–subject binding

- Are the security objectives for the TOE and the TOE environment adequate to counter identified threats or enforce security policies and assumptions?
- Are security requirements internally consistent? Will they lead to the development of a TOE that meets stated security objectives? Are the security requirements explicitly stated, clear, and unambiguous?

Security Target Evaluation, or ASE, is the second security assurance class discussed. This class is invoked after an ST has been developed to determine whether an ST is an adequate, complete, correct, and consistent interpretation of a PP. The activities defined in ASE result in a formal evaluation of an ST.

Exhibit 12. FMT Functional Class: Security Management

Family	Name	Function[20]	Component(s)	Name
FMT_MOF	Management of functions in TSF	allow authorized user roles to control security management functions	FMT_MOF.1	Management of security functions behavior
FMT_MSA	Management of security attributes	allow authorized user roles to control security attributes	FMT_MSA.1	Management of security attributes
			FMT_MSA.2	Secure security attributes
			FMT_MSA.3	Static attribute initialization
FMT_MTD	Management of TSF data	allow authorized user roles to manage TSF data	FMT_MTD.1	Management of TSF data
			FMT_MTD.2	Management of limits on TSF data
			FMT_MTD.3	Secure TSF data
FMT_REV	Revocation	revoke security attributes	FMT_REV.1	Revocation
FMT_SAE	Security attribute expiration	enforce expiration time frames for security attributes	FMT_SAE.1	Time-limited authorization
FMT_SMF*	Specification of management functions	specify the security management functions to be provided by the TOE	FMT_SMF.1	Specification of management functions
FMT_SMR	Security management roles	control the assignment of security management roles to users	FMT_SMR.1	Security roles
			FMT_SMR.2	Restrictions on security roles
			FMT_SMR.3	Assuming roles

* Per Final Interpretation 065.

An ST can be submitted for evaluation prior to or concurrently with a TOE. However, having a formal evaluation of an ST prior to beginning full-scale development of the TOE makes more sense from a cost and schedule perspective; for example, errors and misunderstandings in the ST can be corrected prior to development of the TOE. Evaluation goals, which mirror the APE evaluation goals, are established for the first seven sections of an ST:

- Is the ST identification information an accurate reflection of the ST?
- Is the TOE description coherent, internally consistent, and consistent with the remainder of the ST?
- Is the security environment in which the TOE will operate understood?
- Are the security objectives for the TOE and the TOE environment adequate to counter identified threats or enforce security policies and assumptions?

Exhibit 13. FPR Functional Class: Privacy

Family	Name	Function[20]	Component(s)	Name
FPR_ANO	Anonymity	protect a user identity while a resource or service is used	FPR_ANO.1	Anonymity
			FPR_ANO.2	Anonymity without soliciting information
FPR_PSE	Pseudonymity	ensure that a resource may be used without disclosing a user identity	FPR_PSE.1	Pseudonymity
			FPR_PSE.2	Reversible pseudonymity
			FPR_PSE.3	Alias pseudonymity
FPR_UNL	Unlinkability	ensure that resources or services may be used in multiple instances by the same user without association of this fact	FPR_UNL.1	Unlinkability
FPR_UNO	Unobservability	ensure that resources or services may be used without disclosing which user is using them	FPR_UNO.1	Unobservability
			FPR_UNO.2	Allocation of information impacting unobservability
			FPR_UNO.3	Unobservability without soliciting information
			FPR_UNO.4	Authorized user observability

- Are the security requirements internally consistent? Will they lead to the development of a TOE that meets stated security objectives? Are the security requirements explicitly stated, clear, and unambiguous?
- Have all SFRs been met by security functions? Have all SARs been met by security assurance measures?
- Is the ST a correct instantiation of the PP?

Configuration Management, or ACM, is the first of seven assurance classes discussed that evaluate TOE conformance. ACM enforces a degree of formality on the development process to prevent the accidental or intentional introduction of security vulnerabilities. ACM evaluates the effective use of CM automation tools, specifically the ability to prevent unauthorized modification of TOE security functions during development, operations, and maintenance. ACM examines the ability of CM processes and procedures to ensure that a TOE contains all configuration items and the correct version of each, prior to delivery. ACM investigates the extent of CM tracking systems and data and

Exhibit 14. FPT Functional Class: Protection of the TSF

Family	Name	Function[20]	Component(s)	Name
FPT_AMT	Underlying abstract machine test	define requirements for testing the underlying abstract machine upon which the TSF relies	FPT_AMT.1	Abstract machine testing
FPT_FLS	Fail secure	ensure that TOE failures result in a known secure state	FPT_FLS.1	Failure with preservation of secure state
FPT_ITA	Availability of exported TSF data	prevent loss of availability of TSF data while it is transported between the TSF and a remote trusted IT product	FPT_ITA.1	Inter-TSF availability within a defined availability metric
FPT_ITC	Confidentiality of exported TSF data	ensure confidentiality of TSF data while it is transported between the TSF and a remote trusted IT product	FPT_ITC.1	Inter-TSF confidentiality during transmission
FPT_ITI	Integrity of exported TSF data	ensure integrity of TSF data while it is transported between the TSF and a remote trusted IT product	FPT_ITI.1	Inter-TSF detection of modification
			FPT_ITI.2	Inter-TSF detection and correction of modification
FPT_ITT	Internal TOE TSF data transfer	protect TSF data when it is transported within a TOE	FPT_ITT.1	Basic internal TSF data transfer protection
			FPT_ITT.2	TSF data transfer separation
			FPT_ITT.3	TSF data integrity monitoring
FPT_PHP	TSF physical protection	restrict unauthorized physical access to the TSF; deter and resist unauthorized physical modification of the TSF	FPT_PHP.1	Passive detection of physical attack
			FPT_PHP.2	Notification of physical attack
FPT_RCV	Trusted recovery	determine if the TOE is started up or recovered without protection mechanisms being compromised	FPT_RCV.1	Manual recovery
			FPT_PHP.3	Resistance to physical attack
			FPT_RCV.2	Automated recovery
			FPT_RCV.3	Automated recovery without undue loss
			FPT_RCV.4	Function recovery

Exhibit 14. FPT Functional Class: Protection of the TSF (continued)

Family	Name	Function[20]	Component(s)	Name
FPT_RPL	Replay detection	detect and preempt action in response to replayed messages, requests, etc.	FPT_RPL.1	Replay detection
FPT_RVM	Reference mediation	ensure that TSPs are always invoked and enforced.	FPT_RVM.1	Non-bypassability of the TSP
FPT_SEP	Domain separation	ensure that the execution of the TSF is protected from external interference and tampering.	FPT_SEP.1	TSF domain separation
			FPT_SEP.2	SFP domain separation
			FPT_SEP.3	Complete reference monitor
FPT_SSP	State synchrony protocol	require critical security functions to use a trusted synchronization protocol.	FPT_SSP.1	Simple trusted acknowledgment
			FPT_SSP.2	Mutual trusted acknowledgment
FPT_STM	Time stamps	provide a reliable time stamp function within a TOE.	FPT_STM.1	Reliable time stamps
FPT_TDC	Inter-TSF basic TSF data consistency	ensure the consistent interpretation of TSF data exchanged with a trusted IT product.	FPT_TDC.1	Inter-TSF basic TSF data consistency
FPT_TRC	Internal TOE TSF data replication consistency	ensure consistent replication of TSF data within a TOE.	FPT_TRC.1	Internal TSF consistency
FPT_TST	TSF self-test	define requirements for automatic self-tests of TSF operation.	FPT_TST.1	TSF testing

whether they capture all configuration items, including documentation, problem reports, configuration options, and development tools.

Delivery and Operation, or ADO, is the second assurance class that evaluates TOE conformance. ADO ensures that no security vulnerabilities are introduced during the delivery process by preventing and detecting attempted modifications to the TOE at this time. Likewise, ADO ensures that the TOE has been initialized in a secure manner in the operational environment.

The next class, Development, or ADV, prevents the accidental or intentional introduction of security vulnerabilities during the development process by examining seven key areas. Functional specifications are evaluated to demonstrate that all TOE SFRs have been addressed. High-level designs are evaluated to demonstrate that the proposed security architecture is indeed an appropriate implementation of the SFRs. The actual implementation of the

Exhibit 15. FRU Functional Class: Resource Utilization

Family	Name	Function[20]	Component(s)	Name
FRU_FLT	Fault tolerance	ensure that the TOE continues to operate correctly in the presence of stated failure conditions.	FRU_FLT.1	Degraded fault tolerance
			FRU_FLT.2	Limited fault tolerance
FRU_PRS	Priority of service	assign priorities for TOE resource utilization.	FRU_PRS.1	Limited priority of service
			FRU_PRS.2	Full priority of service
FRU_RSA	Resource allocation	control the use of TOE resources.	FRU_RSA.1	Maximum quotas
			FRU_RSA.2	Minimum and maximum quotas

Exhibit 16. FTA Functional Class: TOE Access

Family	Name	Function[20]	Component(s)	Name
FTA_LSA	Limitation on scope of selectable attributes	limit the extent of user security attributes for a given session	FTA_LSA.1	Limitation on scope of selectable attributes
FTA_MCS	Limitation on multiple concurrent sessions	limit the number of concurrent sessions a given user may have	FTA_MCS.1	Basic limitation on multiple concurrent sessions
			FTA_MCS.2	Per-user attribute limitation on multiple concurrent sessions
FTA_SSL	Session locking	provide capability for session locking, whether initiated by a user or TOE security function	FTA_SSL.1	TSF initiated session locking
			FTA_SSL.2	User-initiated session locking
			FTA_SSL.3	TSF initiated termination
FTA_TAB	TOE access banners	display advisory warning to TOE users	FTA_TAB.1	Default TOE access banners
FTA_TAH	TOE access history	display a user's TOE access history	FTA_TAH.1	TOE access history
FTA_TSE	TOE session establishment	deny session establishment	FTA_TSE.1	TOE session establishment

TOE (source code, logic diagrams, firmware, schematics, and so forth) is evaluated to determine if it is complete and structured. The modularity, structure, cohesiveness, and design complexity of TOE security functions is examined. The low-level design is reviewed to demonstrate if it is an accurate and efficient decomposition of the high-level design. The consistency, cor-

Exhibit 17. FTP Functional Class: Trusted Path/Channels

Family	Name	Function[20]	Component(s)	Name
FTP_ITC	Inter TSF trusted channel	provide trusted communication channel between TOE security functions and other trusted IT products	FTP_ITC.1	Inter=TSF trusted channel
FTP_TRP	Trusted path	provide trusted communication path between users and TOE security functions	FTP_TRP.1	Trusted path

rectness, and completeness of the different levels of abstraction that represent TOE security functions are checked. Finally, ADV ensures that TOE security policies are enforced by SFRs and the security functions that implement them.

The AGD class, Guidance Documents, ensures that system administrators and end users have the information they need to use a TOE in a secure fashion. In particular, administrator guidance is evaluated to determine if accurate, complete, and current information is conveyed to personnel responsible for configuring, maintaining, and operating TOE security functions. User guidance is evaluated to determine if accurate, complete, and current information is conveyed to end users which describes TOE security functions and their intended secure use.

The ALC class, Lifecycle Support, evaluates the effectiveness of lifecycle processes and procedures used by the developer to prevent and detect the accidental or intentional introduction of security vulnerabilities. Four key areas are examined:

1. Do lifecycle processes reduce the potential for physical, procedural, and personnel security threats in the development environment?
2. Are flaw-remediation procedures effective?
3. Is the lifecycle model well-defined, appropriate, and measurable? Is the lifecycle model really being followed?
4. Are the tools and techniques used to develop, analyze, and implement TOE security functions appropriate?

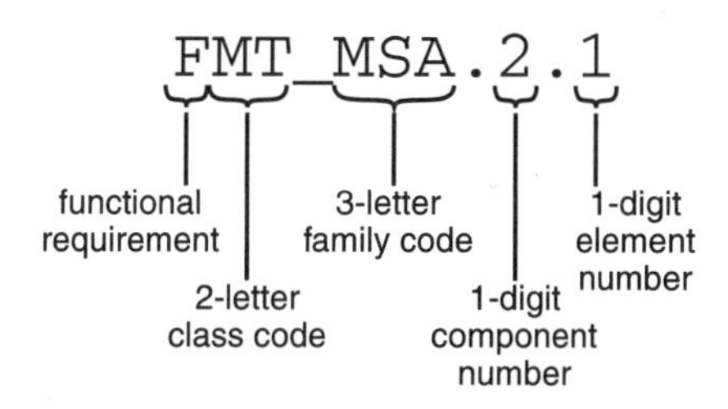

Exhibit 18. Standard Notation for Functional Classes, Families, Components, and Elements

Exhibit 19. Security Assurance Classes

Short Name	Long Name	Type*	Purpose[21]
APE	Protection profile evaluation	PP/ST	demonstrate that the PP is complete, consistent, and technically sound
ASE	Security Target evaluation	PP/ST	demonstrate that the ST is complete, consistent, technically sound, and suitable for use as the basis for a TOE evaluation
ACM	Configuration management	TOE	control the process by which a TOE and its related documentation is developed, refined, and modified
ADO	Delivery and operation	TOE	ensure correct delivery, installation, generation, and initialization of the TOE
ADV	Development	TOE	ensure that the development process is methodical by requiring various levels of specification and design and evaluating the consistency between them
AGD	Guidance documents	TOE	ensure that all relevant aspects of the secure operation and use of the TOE are documented in user and administrator guidance
ALC	Lifecycle support	TOE	ensure that methodical processes are followed during the operations and maintenance phase so that security integrity is not disrupted
ATE	Tests	TOE	ensure adequate test coverage, test depth, and functional and independent testing
AVA	Vulnerability assessment	TOE	analyze the existence of latent vulnerabilities, such as exploitable covert channels; the misuse or incorrect configuration of the TOE; the ability to defeat, bypass, or compromise security credentials
AMA	Maintenance of assurance	AMA	assure that the TOE will continue to meet its Security Target as changes are made to the TOE or its environment

* PP/ST—Protection Profile or Security Target evaluation
TOE—TOE conformance evaluation
AMA— maintenance of assurance after certification.

Tests, represented by the ATE class, are a key component of security assurance. ATE investigates four dimensions of testing. The sufficiency of test coverage (as documented in test plans, test procedures, and test analysis reports) is evaluated to determine if TOE security functions have been adequately exercised. The depth of testing conducted by the developer is examined to ascertain if the structural aspects of TOE security functions have been adequately stressed. The extent of functional testing conducted by the developer is analyzed to confirm its sufficiency. In addition, independent functional and structural testing may be conducted by the evaluator.

Vulnerability Assessments, the AVA class, are a key component of security assurance as well. AVA probes four avenues of potential vulnerabilities. The presence of unintended information flows and the feasibility of exploiting them are assessed. The potential for the TOE to be accidentally or intentionally configured, installed, or operated in an undetectable insecure state is investigated. The robustness and integrity of security mechanisms are analyzed. The extent, severity, and exploitation potential of residual and latent vulnerabilities are discerned.

The AMA class, Maintenance of Assurance, is invoked after a TOE is certified to ensure that security assurance is maintained between certification cycles.

An Assurance Maintenance Plan is created to identify processes the developer must follow to maintain TOE certification during the operations and maintenance phase; evaluators verify its completeness and appropriateness. TOE components are categorized by their relevance to security. This information is used as input to ongoing security impact analysis tasks, which must be performed before any changes are deployed. As tasks and activities in the Assurance Maintenance Plan are performed, evidence of such is collected and organized. Evaluators review this evidence to verify that developers are indeed adhering to their Assurance Maintenance Plan.

An assurance family is a grouping of SARs that share security objectives. The members of a family are referred to as components.[19] Each assurance family is assigned a long name and a three-character mnemonic that is appended to the assurance class mnemonic. Family behavior is described. Unlike functional families, the members of an assurance family only exhibit linear hierarchical relationships, with an increasing emphasis on scope, depth, and rigor. Some families contain application notes that provide additional background information and considerations concerning the use of a family or the information it generates during evaluation activities. Exhibits 20 to 29 list security assurance families by class.

Components are a specific set of security requirements that are constructed from elements; they are the smallest selectable set of elements that can be included in a PP, ST, or package.[19] Components are assigned a long name and described. Hierarchical relationships between one component and another are identified. The short name for a component consists of the class mnemonic, the family mnemonic, and a unique number. Again, application notes may be included to convey additional background information and considerations.

An element is an indivisible security requirement that can be verified by an evaluation and the lowest level security requirement from which components are constructed.[19] One or more elements are stated verbatim for each component. If a component has more than one element, all of them must be used. Dependencies between elements are listed. Elements are the building blocks from which a PP or ST is created. Each assurance element has a unique number that is appended to the component identifier and a one-character code. A "D" indicates assurance actions to be taken by the TOE developer; "C" explains the content and presentation criteria for assurance evidence (i.e., what must be demonstrated);[21] and "E" identifies action to be taken or analyses to be performed by the evaluator to confirm that evidence requirements have been met. Exhibit 30 illustrates the standard notation for assurance classes, families, components, and elements. Annex F provides a glossary of assurance classes and families.

Part 3 of ISO/IEC 15408 also defines seven hierarchical evaluation assurance levels, or EALs. An EAL is a grouping of assurance components that represents a point on the predefined assurance scale.[19,24,110] In short, an EAL is an assurance package. The intent is to ensure that a TOE is not over- or underprotected by balancing the level of assurance against cost, schedule, technical, and mission constraints. Each EAL has a long name and a short name, which consists of "EAL" and a number from 1 to 7. The seven EALs

Exhibit 20. APE Assurance Class: Protection Profile Evaluation

Family	Name	Function[21]	Component(s)	Name
APE_DES	TOE description	determine if the description of the TOE is coherent, internally consistent, and consistent with the rest of the PP	APE_DES.1	TOE description evaluation requirements
APE_ENV	Security environment	determine if the security environment in which the TOE will operate is understood	APE_ENV.1	Security environment evaluation requirements
APE_INT	PP introduction	determine if PP identification information, required for registration, is an accurate reflection of the PP	APE_INT.1	PP introduction evaluation requirements
APE_OBJ	Security objectives	determine if security objectives for the TOE and the environment are adequate to counter identified threats and/or enforce policies and assumptions	APE_OBJ.1	Security objectives evaluation requirements
APE_REQ	IT security requirements	determine if security requirements are internally consistent and will lead to the development of a TOE that meets stated security objectives	APE_REQ.1	IT security requirements evaluation requirements
APE_SRE	Explicitly stated IT security requirements	determine if explicitly stated requirements are clearly and unambiguously stated	APE_SRE.1	Explicitly stated IT security requirements evaluation requirements

add new and higher assurance components as security objectives become more rigorous. Application notes discuss limitations on evaluator actions or the use of information generated. Exhibit 31 cites the seven standard EALs. (EALs are discussed in more detail in Chapter 5.)

2.2.2 The CEM

The Common Methodology for Information Technology Security Evaluation, known as the CEM (or CM), was created to provide concrete guidance to evaluators on how to apply and interpret SARs and their developer actions, content and presentation criteria, and evaluator actions, so that evaluations are consistent and repeatable. To date, the CEM consists of two parts and a supplement. Part 1 of the CEM defines the underlying principles of evaluations and delineates the roles of sponsors, developers, evaluators, and national evaluation authorities. Part 2 of the CEM specifies the evaluation methodology

Exhibit 21. ASE Assurance Class: Security Target Evaluation

Family	Name	Function[21]	Component(s)	Name
ASE_DES	TOE description	determine if the description of the TOE is coherent, internally consistent, and consistent with the rest of the ST	ASE_DES.1	TOE description evaluation requirements
ASE_ENV	Security environment	determine if the security environment in which the TOE will operate is understood	ASE_ENV.1	Security environment evaluation requirements
ASE_INT	ST introduction	determine if ST identification information is an accurate reflection of the ST	ASE_INT.1	ST introduction evaluation requirements
ASE_OBJ	Security objectives	determine if security objectives for the TOE and the environment are adequate to counter identified threats and/or enforce policies and assumptions	ASE_OBJ.1	Security objectives evaluation requirements
ASE_PPC	PP claims	determine if ST is a correct instantiation of the PP	ASE_PPC.1	PP claims evaluation requirements
ASE_REQ	IT security requirements	determine if security requirements are internally consistent and will lead to the development of a TOE that meets stated security objectives	ASE_REQ.1	Information technology (IT) security requirements evaluation requirements
ASE_SRE	Explicitly stated IT security requirements	determine if explicitly stated requirements are clearly and unambiguously stated	ASE_SRE.1	Explicitly stated IT security requirements evaluation requirements
ASE_TSS	TOE summary specification	determine if all SFRs have been met by security functions, and all SARs by assurance measures	ASE_TSS.1	TOE summary specification evaluation requirements

in terms of evaluator tasks, subtasks, activities, subactivities, actions, and work units, all of which tie back to the assurance classes. A supplement was issued to Part 2 in 2002 that provides evaluation guidance for the ALC_FLR family. Like the CC, the CEM will become an ISO/IEC standard in the near future. The CEM is discussed in more detail in Chapter 5.

2.3 Relationship to Other Standards

Like any reasonable standard, the CC/CEM do not, nor are they intended to, operate in a vacuum; rather, extensive interaction occurs between the CC/CEM

Exhibit 22. ACM Assurance Class: Configuration Management

Family	Name	Function[21]	Component(s)	Name
ACM_AUT	CM automation	prevent unauthorized modification of TOE security functions during development, operations, and maintenance	ACM_AUT.1	Partial CM automation
			ACM_AUT.2	Complete CM automation
ACM_CAP	CM capabilities	ensure that TOEs contains all configuration items and the correct version of each prior to delivery	ACM_CAP.1	Version numbers
			ACM_CAP.2	Configuration items
			ACM_CAP.3	Authorization controls
			ACM_CAP.4	Generation support and acceptance procedures
			ACM_CAP.5	Advanced support
ACM_SCP	CM scope	ensure that all configuration items, including documentation, problem reports, options, and development tools, are tracked by the CM system	ACM_SCP.1	TOE CM coverage
			ACM_SCP.2	Problem tracking CM coverage
			ACM_SCP.3	Development tools CM coverage

Exhibit 23. ADO Assurance Class: Delivery and Operation

Family	Name	Function[21]	Component(s)	Name
ADO_DEL	Delivery	prevent and detect modification to the TOE during delivery	ADO_DEL.1	Delivery procedures
			ADO_DEL.2	Detection of modification
			ADO_DEL.3	Prevention of modification
ADO_IGS	Installation, generation, and start-up	ensure that the TOE has been initialized in a secure manner in the operational environment	ADO_IGS.1	Installation, generation, and start-up procedures
			ADO_IGS.2	Generation log

Exhibit 24. ADV Assurance Class: Development

Family	Name	Function[21]	Component(s)	Name
ADV_FSP	Functional specification	demonstrate that all TOE SFRs have been addressed	ADV_FSP.1	Informal functional specification
			ADV_FSP.2	Fully defined external interfaces
			ADV_FSP.3	Semiformal functional specification
			ADV_FSP.4	Formal functional specification
ADV_HLD	High-level design	demonstrate that the security architecture is an appropriate implementation of the TOE SFRs	ADV_HLD.1	Descriptive high-level design
			ADV_HLD.2	Security enforcing high-level design
			ADV_HLD.3	Semiformal high-level design
			ADV_HLD.4	Semiformal high-level explanation
			ADV_HLD.5	Formal high-level design
ADV_IMP	Implementation representation	determine the completeness and structure of the TOE implementation representation, such as source code, logic diagrams, firmware, schematics	ADV_IMP.1	Subset of the implementation of the TSF
			ADV_IMP.2	Implementation of the TSF
			ADV_IMP.3	Structured implementation of the TSF
ADV_INT	TOE security function (TSF) internals	determine the modularity, structure, cohesiveness, and complexity of the TSF design	ADV_INT.1	Modularity
			ADV_INT.2	Reduction of complexity
			ADV_INT.3	Minimization of complexity
ADV_LLD	Low-level design	demonstrate that TSF high-level design has been accurately and efficiently decomposed into a low-level design	ADV_LLD.1	Descriptive low-level design
			ADV_LLD.2	Semiformal low-level design
			ADV_LLD.3	Formal low-level design

Exhibit 24. ADV Assurance Class: Development (continued)

Family	Name	Function[21]	Component(s)	Name
ADV_RCR	Representation correspondence	determine the consistency, correctness, and completeness of the different levels of abstraction which represent the TSF	ADV_RCR.1	Informal correspondence demonstration
			ADV_RCR.2	Semiformal correspondence demonstration
			ADV_RCR.3	Formal correspondence demonstration
ADV_SPM	Security policy modeling	ensure that the SFRs and security functions enforce TOE security policies	ADV_SPM.1	Informal TOE security policy model
			ADV_SPM.2	Semiformal TOE security policy model
			ADV_SPM.3	Formal TOE security policy model

Exhibit 25. AGD Assurance Class: Guidance Documents

Family	Name	Function[21]	Component(s)	Name
AGD_ADM	Administrator guidance	assist personnel responsible for configuring, maintaining, and operating TOE security functions	AGD_ADM.1	Administrator guidance
AGD_USR	User guidance	describe TOE security functions, their intended and secure use, to end-users	AGD_USR.1	User guidance

and other international standards. The nature of this interaction takes four forms, as shown in Exhibit 32:

1. Providing additional guidance on the interpretation and application of the CC/CEM
2. Defining standard practices and conditions CCTLs must adhere to in order to receive and maintain their accreditation
3. Supplementing the CC/CEM by addressing issues that (by design) are not covered
4. Using CC/CEM artifacts in another context

Almost all standards undergo a formal three- to five-year update, reaffirmation, or withdrawal cycle; therefore, it is important for users to stay abreast of new developments to be sure that they have the most current version of each standard; a change in one standard may impact the overall CC/CEM standards framework.

Exhibit 26. ALC Assurance Class: Lifecycle Support

Family	Name	Function[21]	Component(s)	Name
ALC_DVS	Development security	reduce potential physical, procedural, and personnel security threats in the development environment	ALC_DVS.1	Identification of security measures
			ALC_DVS.2	Sufficiency of security measures
ALC_FLR	Flaw remediation	evaluate efficacy of developer's flaw remediation procedures	ALC_FLR.1	Basic flaw remediation
			ALC_FLR.2	Flaw reporting procedures
			ALC_FLR.3	Systematic flaw remediation
ALC_LCD	Lifecycle definition	evaluate the appropriateness, standardization, and measurability of the lifecycle model used by the developer	ALC_LCD.1	Developer-defined lifecycle mode
			ALC_LCD.2	Standardized lifecycle model
			ALC_LCD.3	Measurable lifecycle model
ALC_TAT	Tools and techniques	evaluate appropriateness of the tools and techniques used to develop, analyze, and implement TOE security functions	ALC_TAT.1	Well-defined development tools
			ALC_TAT.2	Compliance with implementation standards
			ALC_TAT.3	Compliance with implementation standards—all parts

Two standards provide additional guidance on the interpretation and application of the CC/CEM. ISO/IEC PDTR 15446[22] is an "informative technical report," not a standard *per se*. ISO/IEC PDTR 15446[22] (1) explains in detail the purpose of PPs and STs and each section in a PP and ST, (2) provides guidance on how to develop PPs and STs, (3) discusses the interaction between PPs and STs, (4) presents a quality checklist for PP/ST developers, and (5) includes several partial examples.

The ISO and IEC councils appointed a Registration Authority (RA) to act on their behalf. Known as the ISO/IEC JTC 1 RA, this organization is authorized to register PPs, functional packages (FPs), and assurance packages (APs) submitted by an applicant. The RA assigns an entry label, which consists of the

Exhibit 27. ATE Assurance Class: Tests

Family	Name	Function[21]	Component(s)	Name
ATE_COV	Coverage	determine sufficiency of test coverage	ATE_COV.1	Evidence of coverage
			ATE_COV.2	Analysis of coverage
			ATE_COV.3	Rigorous analysis of coverage
ATE_DPT	Depth	determine sufficiency of structural testing	ATE_DPT.1	Testing: high-level design
			ATE_DPT.2	Testing: low-level design
			ATE_DPT.3	Testing: implementation representation
ATE_FUN	Functional tests	determine sufficiency of functional testing conducted by developer	ATE_FUN.1	Functional testing
			ATE_FUN.2	Ordered functional testing
ATE_IND	Independent testing	conduct independent functional and structural testing by evaluator	ATE_IND.1	Independent testing—conformance
			ATE_IND.2	Independent testing—sample
			ATE_IND.3	Independent testing—complete

entry type (PP, FP, or AP), the registration year, and registration number. For example, FP-2002-0010 identifies the tenth functional package that was registered in 2002. The PP, FP, or AP submitted by an applicant must conform to the requirements stated in ISO/IEC 15408; however, RAs only validate the structure and consistency of the PP, FP, or AP; they do not validate technical content.[71]

If the information supplied by the applicant passes the structural and consistency checks, the RA lists the entry as "registered"; if not, the entry is listed as "failed validation." If the PP, FP, or AP has been certified by a National Evaluation Authority, it is listed as "certified"; if not, it only listed as "registered." Registration with the ISO/IEC JTC 1 RA is an additional step beyond being part of the PP Registry of a National Evaluation Authority. The intent is to provide the widest possible publicity for a PP, FP, or AP. ISO/IEC 15292(2001-12) explains the procedures for registering a PP, FP, or AP and the responsibilities of the RA and applicant.

Seven standards define standard practices and conditions Common Criteria Testing Laboratories (CCTLs) must adhere to in order to receive and maintain their accreditation. ISO/IEC 17025,[67] which replaced ISO/IEC Guide 25, and, within the European Union, EN45001[60] define the criteria for establishing and demonstrating competency as a testing laboratory. Particular emphasis is placed on producing consistent and repeatable results. IOS/IEC Guide 65[68] and, within the European Union, EN 45011[61] define procedural standards for the national evaluation authorities who accredit CCTLs. This helps to ensure a consistent standard of laboratory accreditation in all countries. The ISO 9000[73] compendium of International Quality Management standards defines a process for generating and maintaining adequate and accurate documentation to support the recommendation of a CCTL for a specific product to be certified (or not).

These standards are supplemented by country-specific standards, such as the National Voluntary Laboratory Accreditation Program Handbook 150 (NVLAP® Handbook 150)[110] and NVLAP® Handbook 150-20[112,113] in the United States. NVLAP® Handbook 150 defines specific procedures, beyond those required by ISO 9000,[73] that all types of laboratories must follow when collecting, analyzing, storing, and reporting evaluation evidence. NVLAP® Handbook 150-20[112,113] is an extension of NVLAP® Handbook 150[110] that levies additional specific requirements on CCTLs.

Exhibit 28. AVA Assurance Class: Vulnerability Assessment

Family	Name	Function[21]	Component(s)	Name
AVA_CCA	Covert channel analysis	determine the presence and viability of unintended information flows	AVA_CCA.1	Covert channel analysis
			AVA_CCA.2	Systematic covert channel analysis
			AVA_CCA.3	Exhaustive covert channel analysis
AVA_MSU	Misuse	investigate the potential for the TOE to be accidentally or intentionally configured, installed, or operated such that an undetectable insecure state results	AVA_MSU.1	Examination of guidance
			AVA_MSU.2	Validation of analysis
			AVA_AMSU.3	Analysis and testing for insecure states
AVA_SOF	Strength of TOE security functions	analyze the robustness and integrity of security mechanisms	AVA_SOF.1	Strength of TOE security function evaluation
AVA_VLA	Vulnerability analysis	determine the extent, severity, and exploitation potential of residual and latent vulnerabilities	AVA_VLA.1	Developer vulnerability analysis
			AVA_VLA.2	Independent vulnerability analysis
			AVA_VLA.3	Moderately resistant
			AVA_VLA.4	Highly resistant

Exhibit 29. AMA Assurance Class: Maintenance of Assurance

Family	Name	Function[21]	Component(s)	Name
AMA_AMP	Assurance maintenance plan	identify processes that developer must follow to maintain TOE certification during operations and maintenance phase	AMA_AMP.1	Assurance maintenance plan
AMA_CAT	TOE component categorization report	categorize TOE components by security relevance	AMA_CAT.1	TOE component categorization report
AMA_EVD	Evidence of assurance maintenance	generate evidence that assurance maintenance plan is being followed	AMA_EVD.1	Evidence of maintenance process
AMA_SIA	Security impact analysis	verify that security impact of proposed changes to the certified TOE were analyzed prior to implementation	AMA_SIA.1	Sampling of security impact analysis
			AMA_SIA.2	Examination of security impact analysis

Nine national and international standards supplement the CC/CEM by addressing issues that (by design) are not covered and are out of scope. ISO/IEC 17799(2000-12), originally issued as BS 7799, covers in detail every aspect of OPSEC and development of the associated policies and procedures. The CC/CEM focuses on specifying security policies that are enforced by functional requirements that implement security objectives and verified by assurance activities. In contrast, ISO/IEC 17799 discusses implementing and maintaining security policies. More emphasis is placed on the human and organizational framework for managing security. Procedures for identifying assets and classifying their sensitivity are explained. Personnel, physical, and environmental security issues are examined as they relate to the development, operations, and maintenance lifecycle phases. Special attention is paid to access control challenges for mobile users as well as contingency planning and disaster recovery. Finally, ISO/IEC 17799 tackles an important topic often avoided by other security standards — compliance with laws, regulations, and

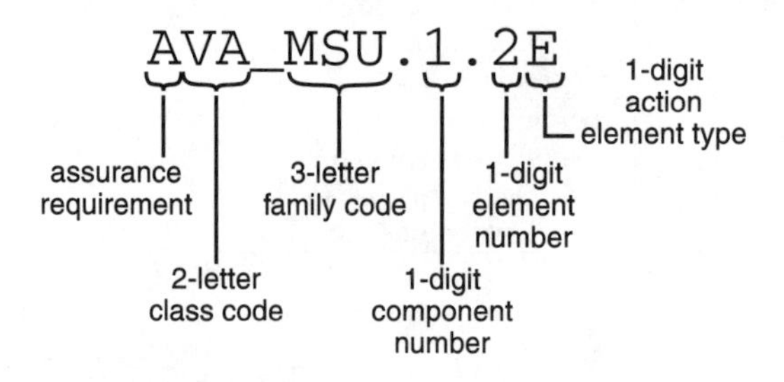

Exhibit 30. Standard Notation for Assurance Classes, Families, Components, and Elements

Exhibit 31. Standard EAL Packages

Short Name	Long Name	Level of Confidence
EAL 1	Functionally tested	Lowest
EAL 2	Structurally tested	
EAL 3	Methodically tested and checked	
EAL 4	Methodically designed, tested, and reviewed	Medium
EAL 5	Semiformally designed and tested	
EAL 6	Semiformally verified design and tested	
EAL 7	Formally verified design and tested	Highest

other legal requirements. The standard is intended to be used as a starting point for an organization to develop its own specific OPSEC policies and procedures. Topics discussed include:

- Information security policy
- Organizational security
  - Internal
  - Interactions with third parties
  - Outsourcing
- Asset classification and control
- Personnel security
  - Staffing
  - Training
  - Reporting and responding to security incidents
- Physical and environment security
- Communications and operations management
- Access control
  - Users
  - Network access control
  - Operating system access control
  - Application system access control
  - Monitoring and logging
  - Remote access
- System development and maintenance
- Business continuity management
- Compliance with laws and regulations

The ISO/IEC TR 13335 five-part standard is also not a standard *per se*, but rather a technical report (TR). The purpose of this series is to provide guidance about the management aspects of IT security, such as:[62–66]

- Concepts associated with the management of IT security
- Relationships between the management of IT security and management of IT in general
- Models that can be used to explain and analyze IT security

Exhibit 32. Relationship of the CC/CEM to Other Standards

Standard	Interaction with CC/CEM
I. Provide Additional CC/CEM Guidance	
ISO/IEC PDTR 15446(2001-4), Information Technology—Security techniques, Guide to the production of protection profiles and security targets	• provides guidance on how to develop PPs and STs; • discusses interaction between PPs and STs
ISO/IEC 15292(2001-12) Information Technology — Security Techniques — Protection Profile registration procedures	• defines responsibilities and procedures for registering a PP, FP, or AP with the ISO/IEC JTC 1 RA.
II. Define Standard CCTL Practices	
ISO/IEC 17025(1999), General Requirements for the competence of calibration and testing laboratories or EN45001, General criteria for the operation of testing laboratories, CEN/CENELEC, 1989	• levies requirements on prospective CCTLs
ISO/IEC Guide 65(1996) General requirements for bodies operating product certification systems, or EN 45011, General criteria for certification bodies operating product certification systems, CEN/CENELEC, 1989	• levies requirements on national evaluation authorities who accredit CCTLs
ISO 9000 Compendium, International Standards for Quality Management	• defines quality management standards that the CCTLs must adhere
*NVLAP® Handbook 150, Procedures and General Requirements, NIST, Department of Commerce, July 2001.	• defines procedures that CCTLs must follow when collecting, analyzing, storing, and reporting evaluation evidence
*NVLAP® Handbook 50-20, Information Security Testing—Common Criteria, version 1.1, NIST, Department of Commerce, April 1999.	• defines procedures that CCTLs must follow when conducting CC/CEM evaluations
III. Supplement CC/CEM	
ISO/IEC 13335-1(1996-12), Information Technology—Guidelines for the management of IT security—Part 1: Concepts and models for IT security	• provides a high-level management overview of the topics, concepts, and models of IT security management
ISO/IEC 13335-2(1997-12), Information Technology—Guidelines for the management of IT security—Part 2: Managing and planning IT security	• discusses the activities, roles, and responsibilities related to planning and managing IT security within an organization
ISO/IEC 13335-3(1998-06), Information Technology—Guidelines for the management of IT security—Part 3: Techniques for the management of IT security	• describes and recommends IT security management techniques
ISO/IEC 13335-4(2000-03), Information Technology—Guidelines for the management of IT security—Part 4: Selection of safeguards	• provides guidance for the selection and implementation of IT security safeguards
ISO/IEC 13335-5(2001-11), Information Technology—Guidelines for the Management of IT Security—Part 5: Management guidance on network security	• provides specific guidance for communications network security
ISO/IEC 17799(2000-12), Information Technology—Code of practice for information security management	• addresses OPSEC issues and development of the associated policies and procedures
*NVLAP® Handbook 150-17, Cryptographic Module Testing, NIST, Department of Commerce, June 2000.	• defines test methods used to assess the robustness of cryptographic modules and their conformance to approved cryptographic standards, such as FIPS PUB 140-2

Exhibit 32. Relationship of the CC/CEM to Other Standards (continued)

Standard	Interaction with CC/CEM
Information Assurance Technical Framework (IATF), v3.0, September 2000.	• identifies "best practices" for components of a security architecture in different operational environments
System Security Engineering Capability Maturity Model (SSE-CMM), v2.0, April 1999.	• defines process and metrics to assess a development organization's maturity relative to security engineering
IV. Use CC/CEM Artifacts	
**NSTISSI #1000, National Information Assurance Certification and Accreditation Process (NIACAP), National Security Telecommunications and Information System Security Committee (NSTISSC), April 2000.	• incorporates results from CC/CEM evaluations, such as ETRs, into the certification and accreditation process

* These standards are applicable in Canada and the U.S.

** This standard is applicable only in the U.S. Equivalent standards are in place in other countries as well.

At the time of writing, the series included five technical reports; more may be added in the future. Information generated from several of the activities described in the series is used as input to a PP.

ISO/IEC TR 13335 Part 1 provides a high-level management overview of the topics, concepts, and models of IT security management. ISO/IEC TR 13335 Part 2 discusses the activities, roles, and responsibilities related to planning and managing IT security within an organization. Unlike ISO/IEC 17799, ISO/IEC TR 13335 Part 2 does not discuss outsourcing or interactions with third-party organizations. ISO/IEC TR 13335 Part 3 describes and recommends IT security management techniques. In particular, techniques for performing risk analyses, selecting and implementing safeguards, translating IT security policies into deployable IT security plans, and ongoing security compliance monitoring are presented. ISO/IEC TR 13335 Part 4 continues the discussion on the selection of safeguards. Additional details are provided concerning: (1) organizational, physical, and IT system-specific safeguards; (2) safeguards for confidentiality; (3) safeguards for integrity; (4) safeguards for availability; and (5) safeguards for accountability, authenticity, and reliability. The Annexes contain examples tied to specific industrial sectors, such as finance and healthcare. ISO/IEC TR 13335 Part 5 further explores IT security management issues specifically related to communications networks.

NVLAP® Handbook 150-17,[111] which is used by Canada and the United States, defines test methods that assess the robustness of cryptographic modules and their conformance to approved cryptographic standards. This is referred to as the Cryptographic Module Validation Program (CMVP). The results of cryptographic testing are used as input to a ST to demonstrate that the strength of function and required level of confidentiality have been achieved. Annex E lists the accredited cryptographic module testing laboratories, at the time of writing, by country.

The Information Assurance Technical Framework (IATF)[88] identifies "best practices" for individual components of a security architecture in a variety of different operational environments. This information is useful when defining security requirements and selecting appropriate countermeasures.

The System Security Engineering Capability Maturity Model (SSE-CMM)[89] was initiated by the U.S. National Security Agency (NSA), Office of the Secretary of Defense (OSD), and Communications Security Establishment (CSE) of Canada in April 1993. ISO/IEC 15408 is primarily an assessment of the functional security of a product (or system). In contrast, SSE-CMM[89] is primarily an assessment of the security engineering processes used to develop a product or system. The intent is to provide a standardized assessment that assists customers to determine the ability of a vendor to perform well on security engineering projects. SSE-CMM[89] was derived from the Systems Engineering Capability Maturity Model (SE-CMM) developed by the Software Engineering Institute (SEI). Additional specialized security engineering needs were added to the model so that it incorporates the best-known security engineering practices.[89] SSE-CMM follows the same philosophy as other CMMs, by identifying key process areas (KPAs) and five increasing capability levels. A potential vendor is rated 0 (lowest) to 5 (highest) in each of the 11 security engineering KPAs. An overall rating is given based on the security engineering KPAs and other organizational and project management factors. The consumer then determines if the vendor's rating is appropriate for specific projects. A low SSE-CMM rating may indicate an inability to achieve EAL 3 or above. The long-range plan is for the SSE-CMM, like the CEM, to become an ISO/IEC standard that complements ISO/IEC 15408.

Each country and government agency defines the process and criteria for certifying critical systems; this is often referred to as certification and accreditation (C&A). C&A takes into account other factors that the CC/CEM do not, such as evaluation in the operational environment. In U.S. NSTISSI #1000,[78] the National Information Assurance Certification and Accreditation Process (NIACAP) fills this role. CC/CEM evaluation results, like Evaluation Technical Reports (ETRs), are inputs to the NIACAP, along with other non-CC/CEM observations and results.

2.4 CC User Community and Stakeholders

The CC user community and stakeholders can be viewed according to two different constructs: (1) generic groups of users, and (2) formal organizational entities that are responsible for overseeing and implementing the CC/CEM worldwide (see Exhibit 33). ISO/IEC 15408-1 defines the CC/CEM generic user community to consist of:

- Consumers
- Developers
- Evaluators

Exhibit 33. Roles and Responsibilities of CC/CEM Stakeholders

Category	Roles and Responsibilities*
I. Generic Users[19,21,22]	
Consumers	• specify requirements • inform developers how IT product will be evaluated • use PP, ST, and TOE evaluation results to compare products
Developers	• respond to consumer's requirements • prove that all requirements have been met
Evaluators	• conduct independent evaluations using standardized criteria
II. Specific Organizations[23]	
Customer or End-user	• specify requirements • inform vendors how IT product will be evaluated • use PP, ST, and TOE evaluation results to compare IT products
IT product vendor	• respond to customer's requirements • prove that all requirements have been met • deliver evidence to sponsor
Sponsor	• contract with CCTL for IT product to be evaluated • deliver evidence to CCTL
Common Criteria Testing Laboratory (CCTL)	• request accreditation from National Evaluation Authority • receive evidence from sponsor • conduct evaluations according to CC/CEM • produce Evaluation Technical Reports • make certification recommendation to National Evaluation Authority
National Evaluation Authority	• define and manage National Evaluation Scheme • accredit CCTLs • monitor CCTL evaluations • issue guidance to CCTLs • issue and recognize CC Certificates • maintain Evaluated Products Lists and PP Registry
Common Criteria Implementation Management Board (CCIMB)	• facilitate consistent interpretation and application of the CC/CEM • oversee National Evaluation Authorities • Render decisions in response to RIs • maintain the CC/CEM • coordinate with ISO/IEC JTC1 SC27 WG3 and CEMEB

* See Chapter 5 for a discussion of assurance maintenance roles and responsibilities.

Consumers are those organizations and individuals interested in acquiring a security solution that meets their specific needs. Consumers state their security functional and assurance requirements in a PP. This mechanism is used to communicate with potential developers by conveying requirements in an implementation-independent manner and information about how a product will be evaluated.

Developers are organizations and individuals who design, build, and sell IT security products. Developers respond to a consumer's PP with an implementation-dependent solution in the form of an ST. In addition, developers prove through the ST that all requirements from the PP have been satisfied, including the specific activities levied on developers by SARs.

Evaluators perform independent evaluations of PPs, STs, and TOEs using the CC/CEM, specifically the evaluator activities stated in SARs. The results are formally documented and distributed to the appropriate entities. Consequently, consumers do not have to rely only on a developer's claims; they are privy to independent assessments from which they can evaluate and compare IT security products. As the standard states:[19]

> The CC is written to ensure that evaluations fulfill the needs of consumers — this is the fundamental purpose and justification for the evaluation process.

The Common Criteria Recognition Agreement (CCRA),[23] signed by 15 countries to date, formally assigns roles and responsibilities to specific organizations:

- Customers or end users
- IT product vendors
- Sponsors
- Common Criteria Testing Laboratories (CCTLs)
- National Evaluation Authorities
- Common Criteria Implementation Management Board (CCIMB)

Customers or end users perform the same role as consumers in the generic model. They specify their security functional and assurance requirements in a PP. By defining an assurance package, they inform developers how the IT product will be evaluated. Finally, they use PP, ST, and TOE evaluation results to compare IT products and determine which best meets their specific needs and will work best in their particular operational environment.

IT product vendors perform the same role as developers in the generic model. They respond to customer requirements by developing an ST and corresponding TOE. In addition, they provide proof that all security functional and assurance requirements specified in the PP have been satisfied by their ST and TOE. This proof and related development documentation are delivered to the sponsor.

A new role introduced by the CCRA is that of the sponsor. A sponsor locates an appropriate CCTL and makes contractual arrangements with the laboratory to conduct an evaluation of an IT product. The sponsor is responsible for delivering the PP, ST, or TOE and related documentation to the CCTL and coordinating any preevaluation activities. A sponsor may represent the customer or the IT product vendor or be a neutral third party, such as a system integrator.

The CCRA divides the generic evaluator role into three hierarchical functions: Common Criteria Testing Laboratories, National Evaluation Authorities, and the Common Criteria Implementation Management Board (CCIMB). CCTLs must meet accreditation standards and are subject to regular audit and oversight activities to ensure that their evaluations conform to the CC/CEM. CCTLs receive the PP, ST, or TOE and the associated documentation from the sponsor. They conduct formal evaluations of the PP, ST, or TOE according to the CC/CEM and the assurance package specified in the PP. If missing, ambiguous, or incorrect information is uncovered during the course of an evaluation, the CCTL issues an Observation Report (OR) to the sponsor requesting clarification. The results are documented in an Evaluation Technical Report (ETR), which is sent to the National Evaluation Authority along with a recommendation that the IT product be certified (or not). Annex D lists the accredited CCTLs, at the time of writing, by country.

Each country that is a signatory to the CCRA has a National Evaluation Authority. The National Evaluation Authority is the focal point for CC activities within its jurisdiction. A National Evaluation Authority may take one of two forms — that of a Certificate Consuming Participant or that of a Certificate Authorizing Participant. A Certificate Consuming Participant recognizes CC Certificates issued by other entities but, at present, does not issue any certificates itself. It is not uncommon for a country to sign on to the CCRA as a Certificate Consuming Participant and then switch to a Certificate Authorizing Participant later, after establishing a national evaluation scheme and accrediting some CCTLs.

A Certificate Authorizing Participant is responsible for defining and managing the evaluation scheme within its jurisdiction. This is the administrative and regulatory framework by which CCTLs are initially accredited and subsequently maintain their accreditation. The National Evaluation Authority issues guidance to CCTLs about standard practices and procedures and monitors evaluation results to ensure their objectivity, repeatability, and conformance to the CC/CEM. The National Evaluation Authority issues official CC Certificates, if they agree with the CCTL recommendation, and recognizes CC Certificates issued by other National Evaluation Authorities. In addition, the National Evaluation Authority maintains the Evaluated Products List and PP Registry for its jurisdiction. Annex C lists the National Evaluation Authority for each CCRA participant, at the time of writing, by country.

The Common Criteria Implementation Management Board (CCIMB) is composed of representatives from each country that is a party to the CCRA. The CCIMB has the ultimate responsibility for facilitating the consistent interpretation and application of the CC/CEM across all CCTLs and National Evaluation Authorities. Accordingly, the CCIMB monitors and oversees the National Evaluation Authorities. The CCIMB renders decisions in response to Requests for Interpretation (RIs). (The RI process is discussed in Section 2.5.) Finally, the CCIMB maintains the current version of the CC/CEM and coordinates with ISO/IEC JTC1 SC27 WG3 and the CEMEB concerning new releases of the CC/CEM and related standards. Exhibit 34 illustrates the interaction among the major CC/CEM stakeholders.

2.5 Future of the CC

As mentioned in Section 2.1, the CC/CEM is the result of a 30-year evolutionary process. The CC/CEM and the processes governing them have been designed so that CC/CEM will continue to evolve and not become obsolete when technology changes, such as the *Orange Book* did. Given that, and the fact that 15 countries have signed the CC Recognition Agreement, the CC/CEM will be with us for the long term. Two near-term events to watch for are the issuance of both the CEM and the SSE-CMM as ISO/IEC standards.

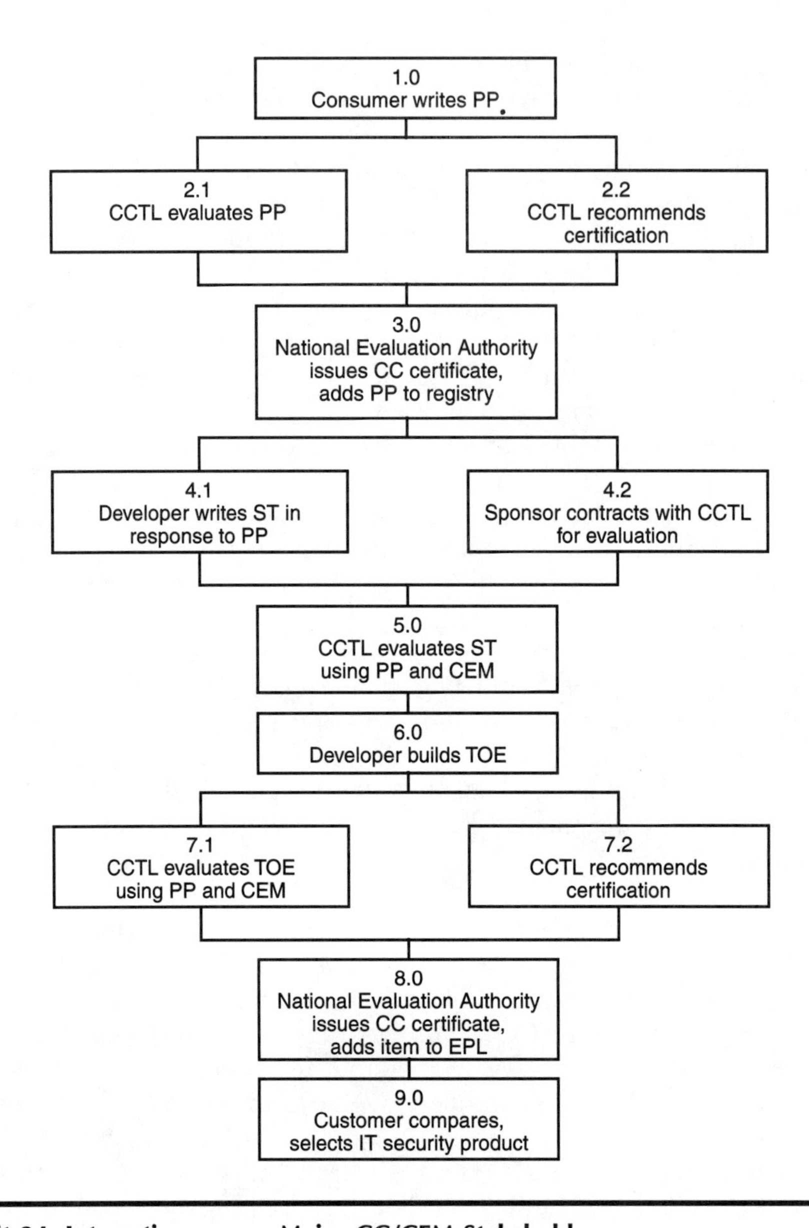

Exhibit 34. Interaction among Major CC/CEM Stakeholders

The Common Criteria Implementation Management Board has set in place a process to ensure consistent interpretations of the CC/CEM and to capture any needed corrections or enhancements to the methodology. Both situations are dealt with through what is known as the Request for Interpretation process (see Exhibit 35). The first step in this process is for a developer, sponsor, or CCTL to formulate a question. This question, or RI, may be triggered by four different scenarios. The organization submitting the RI:[23]

- Perceives an error in the CC or CEM
- Perceives the need for additional material in the CC or CEM

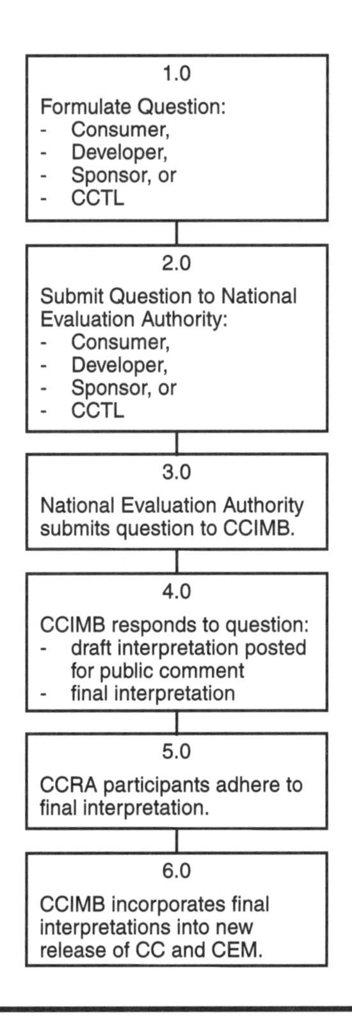

Exhibit 35. RI Process

- Proposes a new application of the CC or CEM and wants this new approach to be validated
- Requests help in understanding part of the CC or CEM

The RI cites the relevant CC or CEM reference and states the problem or question.

Requests for Interpretation are submitted to the National Evaluation Authority, who forwards them to the CCIMB. The CCIMB has a goal of responding to each RI within three months. Initially, a draft interpretation is posted for public comment and review on the Common Criteria Organization Web site.[117] Afterward a final interpretation is posted. At that point, all CCRA participants, National Evaluation Authorities, and CCTLs must adhere to the final interpretation. Finally, final interpretations are incorporated into the CC or CEM. At the time of writing, 33 final interpretations had been issued since the December 1999 release of ISO/IEC 15408. They are summarized in Exhibit 36.

Exhibit 36. CCIMB Final Interpretations*

#	Effective Date	Title of Topic	CC Reference	CME Reference	Summary**
004	11/12/01	ACM_SCP.*.1 C requirements unclear	Part 3ACM_SCP.*.1C ACM_SCP.*.1D	ACM_SCP.*.1C	Clarifies that implementation representation, evaluation evidence, security flaws, and development tools and related documentation are considered part of "list of configuration items," depending on which ACM_SCP.*.1C is invoked.
006	10/15/00	Virtual Machine Description	Part 3 ADV_HLD.*5C		Application note added to ADV_HLD.*.5C equating "underlying hardware, firmware, or software" to virtual machine on which TOE runs.
008	07/31/01	Augmented and Conformant Overlap	Part 1, Section 5.4	ASE_INT	Clarifies difference between conformant and augmented for CC Part 2 and 3.
009	04/13/01	Definition of "counter"	Part 3, page 14, par. 75		Clarifies that: (1) the implementation of objectives counters threats, and (2) countered threats may be mitigated, but not necessarily eradicated.
013	10/15/00	Multiple SOF claims for multiple domains in a single TOE	Part 3 AVA_SOF, ASE_REQ.1.9C		Recognizes validity of multiple SOF domains within a given TOE.
024	01/16/01	Required evaluation evidence for commercial "off the shelf" (COTS) products		Paragraph 34	Explains that assurance requirements apply to entire TOE, including products not under direct control of developer.
025	7/31/01	Level of detail required for hardware descriptions		B.6.2 after paragraph 1817, B.6.3 after paragraph 1818	Explains that the hardware and firmware portions of a TOE must be described at the same level of detail as the software, specifically the impact hardware and firmware features have upon security functions and assurances claimed.
027	02/16/01	Events and functions in AGD_ADM	Part 3 AGD_ADM.1.1C		Clarifies that security-relevant events and administrative functions are not identical.
031	02/16/01	Obvious vulnerabilities	Part 3 AVA_VLA.1 (no change)		Clarifies that: (1) national scheme defines when monitoring of public domain for obvious vulnerabilities should cease, and (2) vulnerabilities found in above time frame that affect ability of TOE to meet stated requirements or counter stated threats must be addressed.

Exhibit 36. CCIMB Final Interpretations* (continued)

#	Effective Date	Title of Topic	CC Reference	CME Reference	Summary**
032	10/15/00	Strength of Function Analysis in ASE_TSS	Part 1, Annex C, Part 3 ASE_TSS		Clarifies that a SOF claim vice analysis must be provided for each function.
033	10/15/00	Use of "check" in Part 3	Part 3 AMA_SIA.*.2E		States that "check" is replaced with "confirm" in AMA_SIA.1.2E and AMA_SIA.2.2E.
037	02/16/01	ACM on Product or TOE?	Part 3 Class ACM		Clarifies that ACM requirements cover the TOE and information related to the TOE, whether the TOE is a product or subset of a product.
043	02/16/01	Meaning of "clearly stated" in APE OBJ.1, ASE OBJ.1	Part 3 APE_OBJ.1 ASE_OBJ.1		States that the term "clearly stated" in APE_OBJ.1.2C, APE_OBJ.1.3C, ASE_OBJ.1.2C, and ASE_OBJ.1.3C is deleted because it duplicates the requirement for coherence in APE_OBJ.1.2E and ASE_OBJ.1.2E.
049	02/16/01	Threats met by environment	Part 1, Annexes B and C.25, Part 3 APE_OBJ.1.3C ASE_OBJ.1.3C		States that the PP/ST description of threats should include threats that are solely countered by measures within the TOE environment.
055	10/15/00	Incorrect Component referenced in Part 2 Annexes, FPT_RCV	Part 2, par. 1236		Corrects typo by referring to FPT_RCV rather than FPT_FLS.1.
058	07/31/01	Confusion over refinement	Part 1, Annex B.26, par. 199b; Annex C.26, par. 215b; Part 2, par. 1		States that changing "the TSF shall" to "the IT environment shall" is a refinement.
062	07/31/01	Confusion over source of flaw reports	Part 3, par. 391 ALC_FLR.2.2D ALC_FLR.3.2D		Clarifies that valid sources of flaw reports include users beyond just TOE users.
064	02/16/01	Apparent higher standard for explicitly stated requirements	Part 3 APE_SRE par. 164, ASE_SRE par. 185	New insert after par. 281 and par. 470	States that existing CC functional and assurance requirements should be used as models of compliance for explicitly stated requirement's measurability and objectivity.
065	07/31/01	No component to call out security function management	Part 2 added new family FMT_SMF		Adds new family to FMT class (SMF) to allow management functions provided by the TOE to be specified.
067	10/15/00	Application notes missing in ST	Part 1 Annex C.2.8		Clarifies that PP and ST application notes are optional.

Exhibit 36. CCIMB Final Interpretations* (continued)

#	Effective Date	Title of Topic	CC Reference	CME Reference	Summary**
069	03/30/01	Informal Security Policy Model		ADV_SPM, new insert after par. 1473 and 1475	Clarifies that the requirement for an ISPM is met by a clear statement of the security policy.
074	10/15/00	Duplicate Informative Text for ATE_COV.2-3 and ATE_DPT.1-3		Par. 1122	States that evaluator guidance for ATE_COV and ATE_DPT is similar but must be taken in context.
075	10/15/00	Duplicate Informative Text for ATE_FUN.1-4 and ATE_IND.2-1		Par. 616, 617, 805, 806, 838, 839, 1143 1144, 1176, 1177, 1602, 1603, 1635, 1636	Unnecessary duplication of informative text is deleted from ATE_IND.*.1 and ATE_FUN.*.4.
080	10/15/00	APE_REQ.1-12 does not use "shall examine ... to determine"		APE_REQ.1-12	Adds term "shall examine" to APE_REQ.1-12.
084	02/16/01	Separate objectives for TOE and Environment		APE_REQ.1-20 ASE_REQ.1-20	Clarifies that APE_REQ.1-20 and ASE_REQ.1-20 include the phrase "are suitable to meet that security objective for the TOE."
092	07/31/01	Release of the TTOE	Part 3 ALC_FLR, insert after paragraph 391		Clarifies that "each release of the TOE" refers to a product or system that is a release of a certified TOE to which changes have been applied (i.e., the current version of the TOE).
094	07/31/01	FLR Guidance Documents Missing	Part 3 ALC_FLR		Adds requirements that specify how users report flaws, how developers receive such reports, and how users register with a developer.
095	02/16/01	ACM_CAP Dependency on ACM_SCP	Part 3 ACM_CAP		Removes dependency of ACM_CAP.3,.4, and.5 on ACM_SCP.1.
116	07/31/01	Indistinguishable Work Units for ADO_DEL		ADO_DEL.*.1 ADO_DEL.*.2, deleted pars. 671, 672, 967, 968, 1341, 1342; new inserts after pars. 670, 966, 1340	Clarifies that no difference exists between necessity and suitability.

Exhibit 36. CCIMB Final Interpretations* (continued)

#	Effective Date	Title of Topic	CC Reference	CME Reference	Summary**
120	07/31/01	Sampling of process expectations unclear		Annex B.2	Clarifies that the evidence needed is that the process is being followed.
127	10/29/01	TSS Work unit not at the right place		ASE_TSS.1-6	States that the work unit is in the proper place and the developer provides the analysis.
128	10/29/01	Coverage of the delivery procedures		ADO_DEL.2-1	States that delivery documentation should cover the entire TOE, but it may contain different procedures for different parts of the TOE.
133	02/16/01	Consistency analysis in AVA_MSU.2		AVA_MSU.2-8, par. 322 deleted	Deletes reference to consistency analysis.

* This list was accurate at the time of writing; see the Common Criteria Web site for new entries.

**This is an unofficial summary; see the Final Interpretation for the complete official explanation.

The ISO/IEC has a five-year reaffirm, update, or withdrawal cycle for standards. This means that the next version of ISO/IEC 15408, which will include all of the final interpretations in effect at that time, should be released near the end of 2004. The CCIMB has indicated that it may issue an interim version of the CC or CEM, prior to the release of the new ISO/IEC 15408 version, if the volume and magnitude of final interpretations warrant such an action. However, the CCIMB makes it clear that it remains dedicated to supporting the ISO/IEC process.[117]

2.6 Summary

The CC represent the culmination of a 30-year saga involving multiple organizations from around the world. The *Orange Book* is often cited as the progenitor of the CC. Actually, the foundation was laid a decade earlier, in January 1973, with the issuance of DoD 5200.28-M, Techniques and Procedures for Implementing, Deactivating, Testing, and Evaluating Secure Resource-Sharing ADP Systems. DoD 5200.28-M was the first standard to define the purpose of security testing and evaluation, the statement of which presaged the goals of the CC. The next historical milestone was the release of the Trusted Computer System Evaluation Criteria (TCSEC), commonly known as the *Orange Book*, in 1983. The *Orange Book* proposed a layered approach for rating the strength of COMPUSEC features. Four evaluation divisions composed of seven classes were defined, from A1, the highest, to D1, the lowest. Evaluation criteria were grouped into four categories: security policy, accountability, assurance, and documentation. Several standards, known as the Rainbow Series, were issued to interpret or expand the *Orange Book*; however, it was difficult to keep them attuned with new technology. Similar developments were underway in Canada and several European countries. In 1993, these countries joined forces to initiate the Common Criteria Project. The end result was the approval of the three-part standard, ISO/IEC 15408, in December 1999.

The Common Criteria provide a complete methodology for specifying IT security requirements, designing a solution to meet those requirements, and conducting an independent evaluation of a product to ensure that all security requirements have been implemented and that they have been implemented correctly. The CC/CEM provides a comprehensive means of communicating IT security requirements, design information, and evaluation results among multiple parties. The goal of the CC project was to develop a methodology that would be widely recognized and yield consistent repeatable results. Once this standard was in place, it was thought, the quantity, quality, and cost-effectiveness of commercially available IT security products would increase and the time to evaluate them would decrease, especially given the emergence of the global economy. The CC uses the term "IT product" quite broadly, to include hardware, software, and firmware for a single product or multiple products configured as a system or network. Administrative security procedures, physical security, and personnel security are out of the scope of the CC.

The three-part CC standard, ISO/IEC 15408, and the CEM are the two major components of the CC methodology. Part 1 of the ISO/IEC 15408 introduces four key concepts: Protection Profiles, Security Targets, targets of evaluation, and packages. PPs are implementation independent, while STs are implementation dependent. TOEs can be component, composite, or monolithic. Part 2 of ISO/IEC 15408 is a standardized catalog of security functional requirements. Part 3 of ISO/IEC 15408 is a standardized catalog of security assurance requirements. SARs define the criteria for evaluating PPs, STs, and TOEs and the security assurance activities and responsibilities of developers and evaluators. In addition, seven hierarchical security assurance packages, called evaluation assurance levels, are defined. SFRs and SARs are organized in a hierarchical structure of classes, families, components, and elements. A standard notation is used to identify security functional and assurance classes, families, components, and elements.

The CEM provides concrete guidance to evaluators on how to apply and interpret SARs and their developer actions, content and presentation criteria, and evaluator actions, so that evaluation results are consistent and repeatable. Evaluator tasks, subtasks, activities, subactivities, and work units are defined, all of which tie back to the assurance classes. The CEM will become an ISO/IEC standard in the near future.

The Common Criteria and Common Evaluation Methodology are not intended to be used in a vacuum; rather, extensive interaction occurs between the CC/CEM and other international standards. The nature of this interaction takes four forms:

1. Providing additional guidance on the interpretation and application of the CC/CEM
2. Defining standard practices and conditions Common Criteria Testing Laboratories must adhere to in order to receive and maintain their accreditation
3. Supplementing the CC/CEM by addressing issues that are (by design) not covered
4. Using CC/CEM artifacts in another context, such as certification and accreditation

ISO/IEC 15408-1 defines the CC/CEM generic user community to consist of consumers, developers, and evaluators. The CC Recognition Agreement formally assigns roles and responsibilities to specific organizations:

- Customers or end users
- IT product vendors
- Sponsors
- Common Criteria Testing Laboratories (CCTLs)
- National Evaluation Authorities
- Common Criteria Implementation Management Board (CCIMB)

The interaction among these stakeholders continues throughout the life of a system.

The CC/CEM will be with us for the long term. Fifteen countries have signed the Common Criteria Recognition Agreement to date, ensuring that the Common Criteria will in fact become the international standard for IT security. The CC/CEM and the processes governing them have been designed so that the CC/CEM will continue to evolve and not become obsolete when technology changes. The CCIMB has established the Request for Interpretation (RI) process to ensure consistent interpretations of the CC/CEM and to capture any needed corrections or enhancements to the methodology.

Next, Chapter 3 demonstrates how to specify security functional and assurance requirements through the development of a Protection Profile.

2.7 Discussion Problems

1. What are the major differences between the CC, the *Orange Book*, and other predecessors to the CC?
2. Who are the major proponents and beneficiaries of the CC, and why?
3. Under what conditions would you not want to use the CC?
4. Describe the relationship and differences between: (a) the CC and the CEM, and (b) the CC and the CMVP.
5. Who should be contacted regarding the meaning of something in the CC or the CEM?
6. Why would an SFR be dependent on an SAR?
7. Explain the standard notation for functional and assurance classes, families, and elements. Note the similarities and differences, if any.
8. What is the purpose of the RI process?
9. Why is the CEM not an ISO/IEC standard?
10. When will the next version of the ISO/IEC 15408 be released?
11. Who owns the CC? Who owns the CEM?
12. Explain the relationship between a TOE and an IT product.
13. Explain the relationship between a TOE and a PP.
14. Who pays for CC evaluations?
15. What is the CCRA?
16. Who issues CC Certificates?
17. Who acknowledges CC Certificates?
18. What is the purpose of an RI?
19. Why are there three types of security assurance classes?
20. Which security functional class addresses the robustness of encryption algorithms?

Chapter 3

Specifying Security Requirements: The Protection Profile

This chapter explains how to express security requirements through the instrument of a Protection Profile (PP) using the Common Criteria (CC) standardized methodology, syntax, and notation. The required content and format of a PP are discussed section by section. The perspective from which to read and interpret PPs is defined. In addition, the purpose, scope, and development of a PP are mapped to both a generic system lifecycle and a generic procurement sequence.

3.0 Purpose

A PP is a formal document that expresses an *implementation-independent* set of security requirements, both functional and assurance, for an information technology (IT) product or system that meets specific consumer needs.[19,23,110] The process of developing a PP guides consumers to elucidate, define, and validate their security requirements, the end result of which is used to: (1) communicate these requirements to potential developers, and (2) provide a foundation from which a Security Target (ST) can be developed and a formal evaluation conducted. As the standard notes:[22]

> The purpose of a PP is to state a security problem rigorously for a given set or collection of systems or products — known as a Target of Evaluation (TOE) — and to specify security requirements to address that problem without dictating how these requirements will be implemented.

Several stakeholders interact with a PP. PPs are written by customers (or end users), read by potential developers (vendors) and system integrators, and reviewed and assessed by evaluators. Furthermore, other customers with similar security requirements may reuse all or part of an existing PP written by another organization. In fact, certified PPs are generally posted on the Web in PP registries maintained by the National Evaluation Authorities to promote sharing of PP "best practices" and minimize the need to "reinvent the wheel." As an example, the National Information Assurance Partnership (NIAP®) in the United States has an ongoing project to acquire PPs for all key technology areas. (Please note that PPs posted on Web sites have often been sanitized to remove corporate proprietary or security sensitive information; hence, they may not be "complete".)

A PP should be written as a stand-alone document; readers should not have to refer back to a multitude of other documents. A PP should present a concise statement of IT security requirements for the relevant system. As noted in Chapter 2, a PP does not normally contain requirements for operational security, personnel security, or physical security. These topics can only be addressed through the use of explicit requirements (see Section 3.6). Like any other requirements specification, IT security requirements contained in a PP should be correct, unambiguous, complete, consistent, ranked for importance and stability, verifiable, modifiable, traceable, and current (see IEEE Standard 830-1999, IEEE Recommended Practice for Software Requirements Specifications). Once written, a PP is not cast in concrete; rather, it is a living document. Updates to a PP may be triggered by:[22]

- Identification of and response to new threats
- Changes in organizational security policies
- Changes in system mission or intended use
- New cost or schedule constraints
- Higher than expected development costs
- Changes in the allocation of requirements between a target of evaluation (TOE) and its environment
- New technology
- Deficiencies uncovered during an evaluation
- (Re)certification activities (Common Criteria or certification and accreditation [C&A])

A variety of system lifecycle models have been developed over the years, such as structured analysis and design, the classic waterfall model, step-wise refinement, spiral development, rapid application development/joint application development (RAD/JAD), object-oriented analysis and design, and other formal methods. While the sequence, duration, and feedback among the phases of each of the models differ, they all contain certain generic lifecycle phases: concept, requirements analysis and specification, design, development, verification, validation, operations and maintenance, and decommissioning. The Common Criteria/Common Evaluation Methodology (CC/CEM) and artifacts are not tied to any specific lifecycle model; rather, they reflect a continuous process of refinement with built in checks. As the standard states, the CC/CEM methodology is:[19]

> ...based on the refinement of the PP security requirements into a TOE Summary Specification expressed in the ST. Each lower level of refinement represents a further decomposition with additional design detail. The least abstract representation is the TOE implementation itself ... The CC requirement is that there should be sufficient design representations presented at sufficient level of granularity to demonstrate: (1) that each refinement level is a complete instantiation of the higher levels (i.e. all TOE security functions, properties, and behavior defined at the higher level of abstraction must be demonstrably present in the lower level), and (2) that each refinement level is an accurate instantiation of the higher levels (i.e. there should be no additional TOE security functions, properties or behavior defined at the lower level of abstraction that are not specified at the higher level).

While the CC/CEM do not dictate a particular lifecycle model, depending on the evaluation assurance level (EAL), the developer may have to justify the model followed. The ALC_LCD security assurance activities evaluate the appropriateness, standardization, and measurability of the lifecycle model used by the developer. ALC_LCD.1, Developer-Defined Lifecycle Model, requires the developer to establish and use a lifecycle model for the development and maintenance of the TOE, including lifecycle documentation. ALC_LCD.2, Standardized Lifecycle Model, adds requirements to explain why the model was chosen, how it is used to develop and maintain the TOE, and how documentation demonstrates compliance with the model. ALC_LCD.3, Measurable Lifecycle Model, adds requirements to explain the metrics used to measure compliance with the lifecycle model during the development and maintenance of the TOE. EAL 4 includes ALC_LCD.1, EALs 5 and 6 include ALC_LCD.2, and EAL 7 includes ALC_LCD.3.

The CC/CEM and artifacts map to generic lifecycle phases. A PP corresponds to the requirements analysis phase — customers state their IT security requirements in PPs, and the quality of these security requirements is verified through the APE class security assurance activities. (See Exhibit 20, Chapter 2.)

Likewise, large system procurements go through a series of generic phases. Pre-award activities include concept definition, feasibility studies, independent cost estimates, the issuance of a request for proposals (RFPs), and proposal evaluation. Post-award activities include contract award; monitoring system development; accepting delivery orders; issuing engineering change orders (ECPs) to correct deficiencies in requirements, design, or development; and, finally, system deployment. After system roll-out and acceptance are complete, organizations generally transition to an operations and maintenance contract that lasts through decommissioning. A PP is part of pre-award procurement activities; PPs are included in the RFP made available to potential offerors. This practice is becoming prevalent among government agencies in the United States, in part because NSTISSP #11, National Information Assurance Acquisition Policy,[75] mandated the use of CC-evaluated IT security products in critical infrastructure systems starting in July 2002. Exhibit 1 aligns CC/CEM artifacts and activities with generic system lifecycle phases and generic procurement phases.

3.1 Structure

A PP is a formal document with specific content, format, and syntax requirements. This formality is imposed to ensure that PPs are accurately and uniformly interpreted by all the different stakeholders. A PP is not written *per se*; rather, it captures the culmination of a series of analyses conducted by customers to elucidate, definitize, and validate their security requirements. As shown in Exhibit 2, a PP consists of seven sections. All sections, except for Section 6 (PP Application Notes), are required. The content and development of each of the seven sections are discussed in detail below. Early on, it was customary to have many pages of front matter prior to Section 1 of the PP that discussed topics such as conventions, terminology, document organization, and so forth. This is an inadvisable practice. Front matter is not binding on a developer, nor is it evaluated by a Common Criteria Testing Laboratory (CCTL). All information about a PP should be placed in the appropriate section where it will be found, read, and evaluated.

Exhibit 1. Mapping of CC/CEM Artifacts to Generic System Lifecycle and Procurement Phases

CC/CEM Artifacts and Activities	Generic System Lifecycle Phases	Generic Procurement Phases
none	Concept	Concept definition Feasibility studies, needs analysis Independent cost estimate
Protection Profile (PP) Security assurance activity: APE	Requirements analysis and specification	Request for proposal (tender) issued by customer
Security Target (ST) Security assurance activity: ASE	Design	Technical and cost proposals submitted by vendors Technical and cost proposals evaluated by customer
Target of Evaluation (TOE) developed by winning vendor Security assurance activities: ACM, ADV	Development	Contract award
Security assurance activities: ATE, AVA	Verification	Acceptance of delivery orders ECPs issued to correct deficiencies in requirements, design, or development
Security assurance activities: ADO, AGD	Validation, installation and checkout	Deployment
Security assurance activities: ALC, AVA, AMA	Operations and maintenance	Transition to maintenance contract
none	Decommissioning	Contract expiration

Exhibit 2. Content of a Protection Profile (PP)

- 1 Protection Profile Information
  - 1.1 Protection Profile Identification
    - 1.1.1 PP Name:
    - 1.1.2 PP Identifier:
    - 1.1.3 Keywords:
    - 1.1.4 EAL:
    - 1.1.5 Common Criteria Conformance Claim and Version:
    - 1.1.6 PP Evaluation Status:
  - 1.2 Protection Profile Overview
    - 1.2.1 PP Overview
    - 1.2.2 Related PPs and Referenced Documents
    - 1.2.3 PP Organization
    - 1.2.4 Acronyms
- 2 TOE Description
  - 2.1 General Functionality
  - 2.2 TOE Boundaries
- 3 TOE Security Environment
  - 3.1 Assumptions
    - 3.1.1 Intended Use
    - 3.1.2 Operational Environment
    - 3.1.3 Connectivity
  - 3.2 Threats
  - 3.3 Organizational Security Policies
- 4 Security Objectives
  - 4.1 Security Objectives for the TOE
  - 4.2 Security Objectives for the Operational Environment
    - 4.2.1 IT Environment
    - 4.2.2 Non-IT Environment
- 5 Security Requirements
  - 5.1 Security Functional Requirements (SFRs)
  - 5.2 Security Assurance Requirements (SARs)
  - 5.3 Security Requirements for the IT Environment
  - 5.4 Security Requirements for the Non-IT Environment
- 6 Application Notes
- 7 PP Rationale
  - 7.1 Security Objectives Rationale
  - 7.2 Security Requirements Rationale

A PP is a cohesive whole; as such, there is extensive interaction among the six required sections. Section 1 introduces a PP by identifying its nature, scope, and status. Section 2 describes the general functionality and boundaries of the TOE and the assets that require protection. Section 3 states the assumptions, analyzes the threats, and cites organizational security policies that are applicable to the TOE security function (TSF). Section 4 delineates security objectives for the TOE and the IT environment. These objectives are derived from an analysis of the assumptions, threats, and security policies articulated in Section 3. Section 5 implements security objectives through a combination of security functional requirements (SFRs) and security assurance requirements (SARs). These SFRs and SARs are derived from an analysis of the sensitivity of the assets to be protected as stated in Section 2 and the perceived risk of compromise presented in Section 3. Section 6, which is optional, provides an opportunity for a customer to relay additional background information to developers and evaluators. The last section, Section 7, proves that requirements specified in Section 5 implement all security objectives stated in Section 4 for the security environment defined in Section 3. This proof is derived from a correlation analysis, and consistency and completeness checks of Section 5 against Sections 3 and 4. Exhibit 3 summarizes the interaction among the sections of a PP.

As a guide, PPs range from 50 to 100 pages in length, with the average distribution of pages per section as follows:

- Section 1. Introduction, 5 percent
- Section 2. TOE Description, 10 percent
- Section 3. TOE Security Environment, 15 percent
- Section 4. Security Objectives, 5 percent
- Section 5. IT Security Requirements, 35 percent
- Section 7. Rationale, 30 percent

3.2 Section 1: Introduction

The first section of a PP, Introduction, is divided into two subsections: PP Identification and PP Overview.

Exhibit 3. Interaction among Sections of a PP

PP Section	Purpose	Source
1. Introduction	identify nature, scope, and status of PP	
2. TOE Description	describe general functionality and boundaries of TOE and the assets that require protection	
3. TOE Security Environment	state assumptions, analyze threats, and cite security policies applicable to the TSF	
4. Security Objectives	Delineate security objectives for the TOE and the IT environment	derived from an analysis of the assumptions, threats, and security policies articulated in Section 3
5. IT Security Requirements	implement security objectives through a combination of security functional requirements SFRs and SARs	derived from an analysis of the sensitivity of the assets to be protected (Section 2) and the perceived risk of compromise (Section 3)
6. PP Application Notes (optional)	provide additional background information	
7. Rationale	demonstrate/prove that specified requirements (Section 5) implement all security objectives (Section 4) in the stated security environment (Section 3)	derived from a correlation analysis, consistency, and completeness checks of Section 5 against Sections 3 and 4

3.2.1 PP Identification

Information provided in the Identification section is used to properly catalog, index, and cross-reference a PP in registries maintained by the local National Evaluation Authority and other Common Criteria Recognition Agreement (CCRA) participants. The first field in the Identification section is the PP name, the second is the PP identifier. On the surface, these two fields appear to be redundant. Actually, the first field is simply the PP (product or system) name, while the second includes the version and date of the PP. This permits multiple versions of a PP for the same system to be entered in a registry. The third field lists keywords associated with the PP, such as technology type, product category, development or user organization, and brand names. The fourth field cites the EAL to which a conformant TOE will be evaluated. Logically, this could be anywhere from EAL 1 to EAL 7. However, at present, very few products or systems are evaluated above EAL 5. The fifth field states the degree to which the PP conforms to the CC standard and to what version of the standard it conforms. In other words, it indicates whether the PP uses standard CC SFRs, SARs, and EALs verbatim or if they have been augmented or extended. The sixth and last field of the Identification section indicates the current evaluation status of the PP.

Exhibit 4 presents two examples of a PP Identification subsection. The first example is for a composite TOE or system that must meet EAL 5. This PP conforms to both Part 2 and Part 3 of the CC. No explicit SFRs have been included, no SARs have been extended, and the EAL has not been augmented. The second example is for a monolithic TOE or commercial "off the shelf" (COTS) product. This PP is conformant to Part 2 of the CC. Part 3, however, has been augmented; hence, the notation that the EAL requirement is "EAL 2 augmented." Neither PP has completed a formal evaluation by an accredited CCTL.

3.2.2 PP Overview

The PP Overview section provides a brief description of the PP and sets the context for the rest of the document. The PP Overview consists of four fields. The first field, which is also called PP Overview, is a stand-alone narrative that summarizes the security problem being solved by the PP. It should be limited to one to two paragraphs. The PP Overview is often used as a stand-alone abstract in PP registries. This information, along with the Identification section, helps a reader determine if the PP may be of interest. The second field lists PPs that are related to this one and any documents referenced in the PP. Related PPs could include PPs in use by the same organization, PPs for systems or products with which this PP must interface, and earlier versions of this PP. Referenced documents may include organizational security standards and policies, national laws and regulations, and Common Criteria publications. The third field, PP Organization, explains the content and structure of the PP.

Exhibit 4. PP Identification Examples

Example 1: Composite TOE for a system

1.1 PP Identification

1.1.1 **PP Name:** High Assurance Remote Access

1.1.2 **PP Identifier:** U.S. DoD Remote Access PP for High Assurance Environments, version 1.0, May 2000

1.1.3 **Keywords:** remote access, network security, remote unit, communications server

1.1.4 **EAL:** EAL 5

1.1.5 **Common Criteria Conformance Claim and Version:** ISO/IEC 15408(12–99), Information Technology — Security Techniques — Criteria for Evaluating IT Security, Part 2 — Conformant, Part 3 — Conformant.

1.1.6 **PP Evaluation Status:** formal evaluation by a CCTL TBD

Example 2: Monolithic TOE for a Commercial COTS Product

1.1 PP Identification

1.1.1 **PP Name:** Medium Assurance Traffic-Filter Firewall

1.1.2 **PP Identifier:** U.S. DoD Traffic-Filter Firewall PP for Medium Robustness Environments, version 1.0, January 2000.

1.1.3 **Keywords:** information flow control, firewall, packet filter, network security

1.1.4 **EAL:** EAL 2 augmented

1.1.5 **Common Criteria Conformance Claim and Version:** ISO/IEC 15408(12–99), Information Technology — Security Techniques — Criteria for Evaluating IT Security, Part 2 — Conformant, Part 3 — Augmented.

1.1.6 **PP Evaluation Status:** informal evaluation complete, formal evaluation by a CCTL TBD.

This is the only "boilerplate" field in a PP. The fourth field defines acronyms as they are used in the PP.

Exhibit 5 provides examples of a PP Overview section. The first example is for a high-assurance remote access system. This overview concentrates on defining the scope of the PP by pointing out that:

- The requirements specified in the PP are the minimum needed.
- Interaction with external entities is needed.
- This is a composite TOE.
- The requirements specified in the PP may not be applicable to all remote access scenarios.

The second example is for an access control product. Four key pieces of information are conveyed:

1. Functions that a compliant product will perform
2. Level of threats a compliant product will protect against (nonhostile inadvertent or casual attempts)
3. Level of risk for the operational environment (moderate)
4. Minimum strength of function (medium)

Exhibit 5. PP Overview Examples

Example 1: System/Composite TOE — High Assurance Remote Access

1.2.1. PP Overview

This Protection Profile specifies the DoD's minimum security needs for remote access connection to a high-assurance enclave. The communications media for remote access may be outside the sphere of ownership and management of the enterprise making the remote connection. The requirements in this PP contain several parameters specified to fit the needs of a particular Remote Access system. Since this PP defines requirements for a system, or composite Target of Evaluation (TOE), which will be implemented through several inter-connected Security Targets. This PP specifies the security policies supported by the TOE and identifies the threats that are to be countered by the TOE. Furthermore, this PP defines implementation-independent security objectives of the system and its environment, defines the functional and assurance requirements, and provides the rationale for the security objectives and requirements. The environment, objectives, and requirements specified within this PP may not be applicable to all remote access scenarios.

Example 2: Product/Monolithic TOE — Medium Assurance Controlled Access

1.2.1. PP Overview

The Controlled Access PP specifies a set of security functional and assurance requirements for IT products that are capable of: (1) enforcing access limitations on individual users and data objects, and (2) providing an audit capability that records the security-relevant events which occur within the system. The Controlled Access PP provides for a level of protection which is appropriate for an assumed non-hostile and well-managed user community requiring protection against threats of inadvertent or causal attempts to breach the system security. The profile is not intended to be applicable to circumstances in which protection is required against determined attempts by hostile and well funded attackers to breach system security. The Controlled Access PP does not fully address the threats posed by malicious system development or administrative personnel. Conformant products are suitable for use in both commercial and government environments. The Controlled Access PP is generally applicable to distributed systems but does not address the security requirements which arise specifically out of the need to distribute the resources within a network. The Controlled Access PP is for a generalized environment with a moderate level of risk to the assets. The assurance requirements (EAL 3) and the minimum strength of function (SOF-medium) were chosen to be consistent with that level of risk.

Example 3: Additional Clause for Government Procurements

1.2.1. PP Overview

Additional security requirements are specified in the Statement of Work (SOW) and Data Item Descriptions (DIDs) regarding security deliverables, achieving and maintaining security certification and accreditation (C&A), and security incident reporting and coordination with the organization's Computer Security Incident Response Center and the Program Office.

Both approaches to writing an overview are acceptable. The first approach is more appropriate for a system-oriented PP; the second approach is more common for a product-oriented PP. The third example contains a clause that government agencies or other organizations may want to add when a PP is part of a procurement. This clause reminds vendors of other contractual security requirements that are not part of the PP.

Exhibit 6 contains an example of the "boilerplate" text that can be used for the PP Organization section. Two variations of the text are possible. First, if the (optional) PP Application Notes are collected in a separate section (Section 6), the Rationale is in Section 7. Second, if the (optional) PP Application Notes are interspersed throughout the PP, the Rationale is in Section 6.

Exhibit 6. PP Organization Example

1.2.3. PP Organization

The main components of the PP are the TOE Description, Security Environment, Security Objectives, IT Security Requirements, and Rationale.

Section 2, TOE Description, provides general information about the functionality of the TOE, defines the TOE boundaries, and provides the context for the PP evaluation.

Section 3, Security Environment, describes aspects of the environment in which the system is to be used and the manner in which it is to be employed. The security environment includes: a) assumptions, b) potential threats, and c) organizational security policies.

Section 4, Security Objectives, identifies security objectives for the TOE and its environment that uphold assumptions, counter potential threats, and enforce organizational security policies.

Section 5, IT Security Requirements, specifies detailed security requirements for the TOE and the operational environment. IT security requirements are subdivided into: (1) security functional requirements that must be implemented, and (2) security assurance requirements that verify the integrity of functional security requirements as implemented.

Section 6, PP Application Notes, contains additional informative material.

Section 7, Rationale, presents evidence that the PP is a complete and cohesive set of IT security requirements and that a conformant TOE would effectively address security needs. The Rationale is organized in two parts. First, a Security Objectives Rationale demonstrates that the stated security objectives counter potential threats. Second, a Security Requirements Rationale demonstrates that: (1) security functional requirements are traceable to security objectives and suitable to meet them, and (2) the specified EAL is appropriate.

Exhibit 7 compares the identifying information captured by a CCRA participant PP registry and the ISO/IEC JTC 1 Registration Authority (RA). As noted in the exhibit, the only three common fields are:

1. 1.1.5 of a PP, CC Conformance Claim and Version, corresponds to 11 of the RA entry.
2. 1.1.6 of a PP, PP Evaluation Status, corresponds to 4 of the RA entry. The RA entry has seven valid status codes from which to choose. In contrast, the PP field is generally limited to whether or not a formal CCTL evaluation has occurred.
3. 1.2.1 of a PP, PP Overview, corresponds to 7 of the RA entry.

In addition, 9 of the RA entry contains the entire PP, functional package (FP), or assurance package (AP). Registries maintained by CCRA participants contain certified PPs.

The information captured by the ISO/IEC JTC 1 RA presents a more complete history of the PP generation and evaluation activities, including its predecessors and successors. In addition, ownership information is recorded. The RA itself originates the majority of the information captured by the RA entry. In comparison, information contained in the CCRA PP Introduction is limited to a single PP and the customer provides it. Also, CCRA Registries do not contain FPs or APs. Both the CCRA PP Registries and the ISO/IEC JTC 1 RA exist for the purpose of technology transfer — disseminating information about PPs so they do not have to be continuously reinvented. It is hoped

Exhibit 7. Comparison of Information Captured by CCRA PP Registries and the ISO/IEC JTC 1 Registration Authority

CCRA PP Registry[19]		ISO/IEC JTC 1 RA[71]	
1.	PP Introduction	1.	Entry Label
1.1	PP Identification	1.1	Entry Type: PP, FP, or AP
1.1.1	PP Name	1.2	Registration Year
1.1.2	PP Identifier	1.3	Registration Number
1.1.3	Keywords	2.	New or Replacement Entry
1.1.4	EAL	2.1	Entry labels of entries replaced by this label
1.1.5***	CC Conformance Claim and Version	2.2	Entry label of any entry replacing this label
1.1.6*	PP Evaluation Status	3.	Draft or Complete Entry
1.2	Protection Profile Overview	4.*	Status of Entry in validation, failed validation, registered, evaluated, certified, obsolescent, retired
1.2.1**	PP Overview	5.	Chronology
1.2.2	Related PPs and Referenced Documents	5.1	Date of Original Acceptance
1.2.3	PP Organization	5.2	Date of last change
1.2.4	Acronyms	5.3	Date of next routine review
		6.	Contact Information
		6.1	Current Sponsor
		6.2	Original Registration Applicant
		6.3	CCTL (for evaluated and certified entries)
		7.**	Executive Summary
		8.	PP Language (if not English)
		9.	Technical Definition (entire PP or package)
		10.	Defect Reports and Resolution Dates
		11.***	Version of CC, CEM, and other CC publications against which the entry was validated by the RA

* — corresponding fields

** — corresponding fields

*** — corresponding fields

that, over time, the identifying information captured by both types of organizations will become more harmonized.

3.3 Section 2: TOE Description

The second required section of a PP is the TOE Description, which contains two subsections: General Functionality and TOE Boundaries.

3.3.1 General Functionality

This subsection describes the general functionality of the product or system, the intended use and the intended operational environment. This subsection

Exhibit 8. TOE Description Examples

Example 1: System/Composite TOE — Switches and Routers

2.1. General Functionality

The TOE for this PP is a switch (voice, Frame Relay, ATM or optical) or router, including all resident cards, ports, software, data, and interfaces. All circuits associated with the switches and routers are also part of the TOE, including the management link. The network management system is not part of the TOE, nor are any network elements that may be connected to the switch or router, such as digital transport (cross-connect) systems, optical transport systems, and encryption devices. However, the TOE must be able to support encryption or interface to an encryption device. The TOE is intended to protect the network management and control functions and allow the reliable transmission of user data within specified performance parameters.

Example 2: System/Composite TOE — Wide Area Network (WAN)

2.1. General Functionality

The TOE for this PP is a wide area network (WAN) which will provide integrated voice, data, and video telecommunications services CONUS-wide. Because this is a WAN, the TOE is limited to layers 1–3 in the ISO/OSI Reference Model. Telecommunications requirements are expressed in terms of service classes and service interfaces. A service class is determined by the following set of parameters: RMA category, latency level, security level, call set-up time limit, call blocking limit, in-band signaling compatibility, modem compatibility, and voice quality. The following service interfaces are defined: analog, switched analog, low speed digital (DDS, RS-232, RS-449, EIA 530, V.35, X.21), high-speed digital (T-1, ISDN PRI, T-3), remote access interfaces (PPP and SUP), Ethernet, FDDI, X.25, IP and DDC. A variety of site types are supported, including air to ground (A/G) communications sites, radar sites, navigation aid sites, weather sites, air route traffic control centers (ARTCC), air traffic control towers (ATCT), automated flight service stations (AFSS), and terminal radar approach controls (TRACON). In total, approximately 600 major communication nodes and 4000 (manned and unmanned) access points will be supported. Four types of operational data are transported: air to ground voice, air to ground data, ground to ground voice, and ground to ground data.

conveys information about the size, scope, and nature of the TOE to potential developers, along with any other relevant domain knowledge. It is important to note that this subsection describes the functionality of the TOE, not the TSF. Exhibit 8 provides two examples of the description of TOE functionality. This information is usually limited to being a few paragraphs to a few pages.

A key component of this subsection is the determination of the assets to be protected and the sensitivity of each. The assets of a system and their value determine the criticality of a system; hence, they should be identified carefully. The threat assessment in Section 3 of a PP uses this information to determine the severity of the consequences if these assets are lost, misused, misappropriated, corrupted, or compromised. The requirements stated in Section 5 of a PP use this information to determine the type and level of protection needed.

Identification of assets has two steps. First, TOE assets are ascertained along with their owner, origin, and security classification. This information is captured in tabular format. The assets are listed and sorted into three categories:

1. *TOE operational data:* The types of operational or mission-critical data that the TOE generates, processes, stores, or transports. Expressed another way, this information is the reason the system exists. Asset data can be aggregated to major types or decomposed to lower level

subtypes, as appropriate; if aggregated, the highest security classification applies.

2. *TOE hardware, software, and firmware:* The major types of hardware, software, and firmware composing the TOE, in particular the TSF. Again, this information can be aggregated or decomposed, as appropriate.
3. *Operational data and documentation:* The major types of information and documentation that are used to operate and maintain the TOE, in particular TSF.

The middle two columns of the table are filled in by identifying the origin and owner of each asset. In some cases, the origin and owner are the same; in other cases, they are not. The origin of an asset may or may not be within the TOE scope of control. The last column captures the security classification or sensitivity of each asset, such as Top Secret, Secret, Confidential, For Official Use Only, Sensitive But Unclassified, and Corporate Proprietary. Exhibit 9 illustrates this step using the wide area network (WAN) example.

Second, the interactions (permitted and not) between users (subjects) and assets (objects) are delineated. To do this, the various categories of users are designated. In general, users are considered to be outside the TOE; they interact with a TOE through the TSF interface. The CC/CEM acknowledges two main categories of users:

1. *Human users:* authorized local or remote end users and authorized system administrators
2. *External IT entities:* processes that act on behalf of a human user or an external TOE

Once the different categories of TOE users have been identified, their access control rights and privileges are defined for each TOE asset. This information is captured in tabular format. The first column lists the assets and is identical to that in the asset sensitivity table. The remaining columns define the access control rights and privileges for each user category. Access control rights define which assets a user category can access. Access control privileges define what functions or operations a user category can perform using that asset. The range of possible access control privileges varies somewhat from PP to PP. Examples of generic access control privileges include read, write, edit, delete, copy, forward, create, execute, and install. It is unlikely that the customer will have all the information about user groups and sub-user groups at the time a PP is written; many details will be added during design and development. As a result, the information contained in this table is not considered cast in concrete. Rather it is presented to give the developer a reasonable idea of the types of access control rights and privileges the deployed TOE must support and as such control development costs. Exhibit 10 illustrates this step using the WAN example.

It is crucial to have a variety of different stakeholders from the customer organization involved in the generation and validation of the tables presented

Exhibit 9. Asset Identification: Step 1

Table x Asset Types and Sensitivities

Asset Type	Asset Origin	Asset Owner	Asset Security Classification/ Sensitivity
I. Data Transported by the Target of Evaluation (TOE)			
1.1 Air to ground voice	FAA, aircraft	Government	SBU
1.2 Air to ground data	FAA, aircraft	Government	SBU
1.3 Ground to ground voice	FAA, NWS, DoD, public, airlines	Government	SBU
1.4 Ground to ground data	FAA, NWS, DoD, public, airlines	Government	SBU
II. TOE Hardware, Software, Firmware			
2.1 Cryptographic keys	FAA, FAA contractors	Government	SSI
2.2 Cryptographic equipment	FAA, FAA contractors	Government, Contractor	SSI
2.3 Telecommunications infrastructure	FAA contractors	Contractor	NR
2.4 Security management hardware, software, firmware	FAA contractors	Contractor	FOUO/SSI
III. Operational Data and Documentation			
3.1 Personnel access lists and clearances	FAA, FAA contractors	Government, contractor	SSI
3.2 Security incident reports and statistics	FAA, FAA contractors	Government	SSI
3.3 Information system security plan	FAA, FAA contractor	Government	FOUO/SSI
3.4 Vulnerability, threat, and risk assessments	FAA, FAA contractor	Government	FOUO/SSI
3.5 Security testing and evaluation plans, procedures, and results	FAA, FAA contractor	Government	FOUO/SSI
3.6 Security configuration and management information	FAA contractor	Contractor	SSI
3.7 Security Target	FAA contractor	Government	FOUO/SSI
3.8 Contingency and Disaster Recovery Plan	FAA, FAA contractor	Government, Contractor	FOUO/SSI

Key: NR—not rated, public information, SBU—sensitive but unclassified,

SSI— security-sensitive information, FOUO—for official use only

in Exhibits 10 and 11 because of the different perspectives they bring to the problem.

3.3.2 TOE Boundaries

One of the first steps in defining security requirements is to define the boundaries of a system. What constitutes a system, however, is relative to one's vantage point; what one person/organization considers a system, another person/organization may consider a subsystem or a collection of systems. Abstractions about systems and their constituent parts can go very high or very low, depending on one's perspective and the purpose of the abstraction.[99]

Exhibit 10. Asset Identification: Step 2

Table x Access Control Rights and Privileges

Asset Type	End-Users	TOE Operational Staff	TSF Operational Staff	Vendor Maintenance Technicians
I. Data Transported by the TOE				
1.1 Air to ground voice	R, W	none	none	none
1.2 Air to ground data	R, D, CO, F	none	none	none
1.3 Ground to ground voice	R, W	none	none	none
1.4 Ground to ground data	R, W, ED, D, CO, F	none	none	none
II. TOE Hardware, Software, Firmware				
2.1 Cryptographic keys	none	none	CO, F, EX	none
2.2 Cryptographic equipment	none	none	EX	none
2.3 Telecommunications infrastructure hardware, software, firmware	EX	R, W, ED, D, CR, CO, F, EX, IN	R, W, ED, D, CR, CO, F, EX, IN	R, EX, IN
2.4 Security management hardware, software, firmware	none	none	R, W, ED, D, CR, CO, F, EX, IN	R, EX, IN
III. Operational Data and Documentation				
3.1 Personnel access lists and clearances	none	R	R, W	none
3.2 Security incident reports and statistics	none	none	R, W, CR, ED, CO, F	none
3.3 Information system security plan	none	none	R, W, CR, ED, F, EX	none
3.4 Vulnerability, threat, and risk assessments	none	none	R, W, CR, ED, CO, F	none
3.5 Security testing and evaluation plans, procedures, and results	none	none	R, W, CR, ED, CO, F	none
3.6 Security configuration and management information	none	none	R, W, CR, ED, D, CO, F, IN, EX	none
3.7 Security Target	none	none	R, W, CR, ED, CO, F	none
3.8 Contingency and Disaster Recovery Plan	none	EX	R, W, CR, ED, CO, F, EX	EX

Key: R: read or listen (view/hear data, run canned and *ad hoc* reports, download files)
W: write or speak (enter information)
ED: edit (modify existing information)
D: delete (mark a file or record for deletion; do not actually erase it, retain for an audit trail)
CR: create (new record, file, report)
CO: copy (information to local workstation, backup repository, or archive)
F: forward (send information to another user)
EX: execute (system software/firmware, BITE, etc.)
IN: install or upgrade (commercial "off the shelf" [COTS] hardware or software)
None: no access

Exhibit 11. TOE Boundary Definition Example

2.2. TOE Boundaries

This PP is for a composite TOE which consists of three component TOEs (see Figure *X* [*Exhibit 12*]).

- **Telecommunications Services (TS):** wide area network telecommunications infrastructure which interfaces to customer premises equipment in each facility. The TS component TOE is composed of two functional packages (FPs): Data Integrity and Data Availability.
- **Network Management (NM):** Systems that manage the configuration, operation, performance, and maintenance of the TS TOE and maintain and report trouble ticket, performance, and outage information. The NM component TOE is composed on four FPs: Network Configuration Management, Performance Monitoring, VPN Management, and Extranet Management.
- **Security Management (SM):** systems that implement, manage, and monitor security for the TS and NM TOEs. The SM component TOE is composed of seven FPs: Authentication, Access Control, Remote Access Control, Credential Management, Firewall Management, Security Monitoring and Reporting, and Encryption Services.

Security requirements are applicable to each of the TOEs, as explained in Section 5 of the PP.

Using the CC/CEM, a TOE can be monolithic, component, or composite; in addition, a TOE can be localized or distributed. A single PP is written for all three types of TOEs. A composite TOE consists of two or more component TOEs. Furthermore, a composite TOE can be composed of multiple composite TOEs. A monolithic or component TOE is equivalent to a product or subsystem. A composite TOE is equivalent to a system. It is essential to ensure that the component TOEs comprising a composite TOE are complete and consistent, especially in regard to inter-TOE functions and operations and interactions with TOE entities. EALs are specified at the composite level to ensure uniform security assurance activities.[22] (This concept is discussed further in Chapter 5.)

TOE entities can be active or passive. An active entity (subject) is the cause of actions that occur internal to a TOE and cause three types of operations to be performed on the information: (1) operations acting on behalf of an authorized user, (2) operations acting on behalf of multiple authorized users, and (3) operations acting on behalf of the TOE itself.[19] A passive entity (object) is the container from which information originates or to which it is stored — the target of operations performed by subjects.[19]

The definition of TOE boundaries has several important ramifications downstream; hence, this decision should not be made lightly. For example, TOE boundary definitions determine the scope of an evaluation (what is included or excluded) by a CCTL, during the initial certification and any future re-certifications. TOE boundary definitions directly impact the time, difficulty, and cost to certify a TOE. Complexity complicates verification in a geometric manner. Consequently, it is preferable to construct a composite TOE with several well-defined, self-contained component TOEs that represent a logical grouping of functions and SFRs than one large, overly complex component TOE. In this way, security assurance and verification activities can be conducted in an incremental manner.

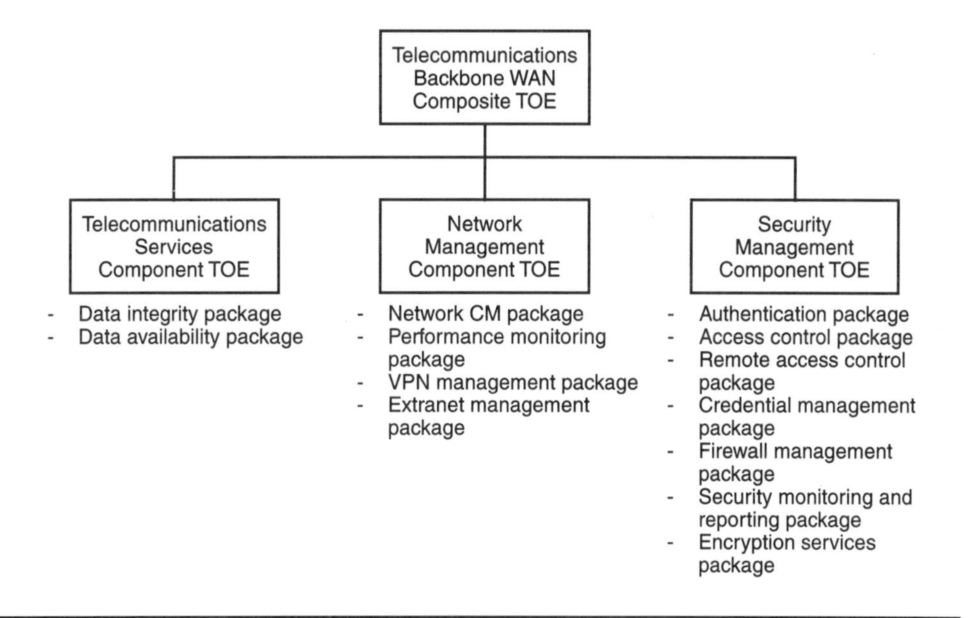

Exhibit 12. TOE Boundary Definition Example

Subsection 2.2 of a PP delineates TOE boundaries. The text should state whether the TOE is monolithic or composite. If composite, all component TOEs should be defined. Functional packages can also be identified at this time. It is important to clarify what is and is not inside the TOE boundaries. The text is generally supplemented with a diagram illustrating the TOE boundaries, as shown in Exhibits 11 and 12.

3.4 Section 3: TOE Security Environment

The third required section of a PP, the TOE Security Environment, defines the nature and scope of TOE security. Three subsections, Assumptions, Threats, and Organizational Security Policies, describe the TOE Security Environment. If the TOE is a distributed system, it may be appropriate to have a separate TOE Security Environment section (3.5.*x*) for each logical partition. Individual assumptions, threats, and policies are numbered (A*x*, T*x*, and P*x*) to permit the consistency and completeness mapping required in Section 7 of the PP (Rationale).

3.4.1 Assumptions

The Assumptions subsection relays pertinent domain knowledge to developers to help them understand the overall framework of the TOE. As McEvilley points out:[101]

> Assumptions are "givens" regarding secure usage of the TOE, scope and boundary of the TOE, and placement of the TOE in its environment, i.e. interaction with other IT and/or people … Assumptions establish the context for all that follows in the PP and ST.

Assumptions about the intended use, operational environment, connectivity, roles, and responsibilities are articulated. Any environmental constraints or operational limitations are clarified. Assumptions cannot be used to mitigate threats. In summary, as the standard notes this subsection is:[19]

> …a statement of assumptions which are to be met by the environment of the TOE in order for the TOE to be considered secure. This statement can be accepted as axiomatic for the TOE evaluation.

Exhibit 13 presents sample Assumptions.

3.4.2 Threats

This subsection characterizes potential threats to the assets identified in Section 2 of the PP, against which protection is required. The CC/CEM does not define how to conduct a threat assessment; rather, it relies on the use of one of the complementary international consensus standards discussed in Section 2.3 of this book. The threat assessment encompasses accidental and malicious intentional attempts to bypass, disable, and compromise security features, functions, and procedures. The TOE, TSF, IT environment, non-IT environment, physical security, personnel security, and operational security are all within the scope of the threat assessment.[101] A comprehensive meaningful threat assessment is needed, for as the standard states:[22]

> The importance of the threat assessment should not be underestimated, since if it is not done properly the TOE may provide inadequate protection, as a result of which the organization's assets may be exposed to an unacceptable level of risk.

Preparing a threat assessment for a PP requires two steps. First, all potential threats are ascertained and itemized. At a high level, potential threats to assets fall into two categories for all PPs:

1. Accidental or malicious intentional compromise of information confidentiality, integrity, and availability by insiders or outsiders
2. Accidental or malicious intentional interruptions to operations due to failures of hardware, software, communication links, power supplies, storage media, and so forth

Exhibit 13. PP Assumptions Example

3.1 Assumptions

This subsection states the assumptions that were made when defining SFRs and SARs. Four types of assumptions are expressed: intended use, operational environment, connectivity, and personnel roles and responsibilities.

3.1.1 Intended Use

- A1 TOE components rely on an underlying operating system and firmware assumed to be installed and operated in a secure manner and in accordance with the ST and other relevant documentation and procedures.
- A2 All TOE files are assumed to be protected from unauthorized access by the underlying operating system.
- A3 The TOE, including the TSF, will meet specified RMA requirements.
- A4 The TOE only consists of the assets described in Section 2 of this PP and only processes data of the sensitivities indicated therein.
- A5 Audit information is reviewed and analyzed on a periodic basis in accordance with the network security policy.
- A6 Cryptographic methods will be resistant to cryptanalytic attacks and will be of adequate robustness to protect sensitive data.

3.1.2 Operational Environment

- A7 All assets are located within controlled access facilities that prevent unauthorized physical access by outsiders. The TOE is installed so that it is protected from casual contact by insiders.
- A8 All TOE hardware, software, and firmware critical to the TOE security function (TSF) are protected from unauthorized modification by hostile insiders and outsiders.
- A9 All equipment complies with environmental standards to ensure physical protection and electrical safety against natural disasters. The TOE has adequate backup power sources to ensure that the sudden loss of power does not affect the availability of services or the loss of data.
- A10 Backup data repositories and archives are located in a secure off-site facility with environmental controls sufficient to ensure data integrity for two years. Chain of custody rules for evidence and evidence preservation are enforced throughout this time interval.
- A11 Network resources are connected to a reliable time source. This will ensure proper synchronization of transmissions among resources and reliable time stamps for auditing network traffic, network performance, network management activities, and security management activities. There is a secondary back-up time source.

3.1.3 Connectivity

- A12 Table x defines the totality of assets for which connectivity will be provided.
- A13 Table y defines the access control rights and privileges by which connectivity will be provided.
- A14 No connectivity or access will be provided that is not defined in Tables x and y.
- A15 Connectivity to trusted and untrusted resources external to the TOE is strictly controlled.

3.1.4 Personnel Roles and Responsibilities

- A16 Only authorized users can access the system and data.
- A17 All authorized users identified in Table y are competent to protect TOE security and the sensitivity of information processed and/or transmitted.
- A18 A variety of users will have access to assets for different reasons and with a different need-to-know. Accordingly, access control rights and privileges will vary by type of user, as defined in Table y.
- A19 Potential attackers are assumed to be insiders or outsiders who have a medium to high level of expertise, resources, and motivation.
- A20 All personnel are properly trained to develop, install, configure and maintain the TOE and the TSF. All personnel follow documented processes and procedures.

High-level threats are decomposed into constituent threats to a level that is meaningful for the PP. Then threats are assigned to applicable component TOEs or FPs. Exhibit 14 illustrates this step using a composite TOE.

Second, the likelihood of each threat occurring is estimated and the severity of the consequences should the threat be instantiated is determined. The severity of the consequences or damage can be cyber or physical, especially in systems where a security compromise can have safety implications, such as an air traffic control system.[99] Because all threats are not equivalent, a risk mitigation priority is established for each potential threat predicated on its severity and likelihood. This approach facilitates prioritizing risk mitigation activities and countermeasures so that resources can be applied to the most critical areas. The severity of the consequences may be expressed as a range, given that instantiation of a particular threat may produce a variety of plausible outcomes (best case to worst case). In this event, the risk mitigation priority is usually expressed as a range as well. Both severity and likelihood are expressed using standardized categories, with likelihood being relative or qualitative as opposed to a precise quantitative measure. Exhibit 15 illustrates this step using a composite TOE.

3.4.3 Organizational Security Policies

This subsection cites Organizational Security Policies (OSPs) that are relevant to the TOE or the TOE environment. OSPs include rules, procedures, and practices that an organization imposes on an IT system to protect its assets.[19] OSPs provide guidance to developers and assist in the formulation of security objectives in Section 4 of a PP. Examples of OSPs include:[22]

- Information flow control rules
- Access control rules
- Policies regarding the use of encryption, such as the requirement for U.S. government agencies to protect sensitive but unclassified (SBU) information with cryptographic modules that are compliant with FIPS 140-2
- Security audit policies and procedures
- Policies regarding use of a standardized IT base

Organizational security policies are unique to each organization and its mission and assets. Local, national, or international laws and regulations may impose additional OSPs — for example, privacy requirements. Accordingly, it is useful to cite the source of OSPs. In general, OSPs fall into seven broad categories, each of which contributes to an overall defense in depth strategy:

1. *Access control* — Access control policies dictate protection from unauthorized access to information and other IT resources.

Exhibit 14. Threat Assessment: Step 1

Table x Potential Threats to Assets by TOE

#	Threat	TOE A	TOE B	TOE C
T1	An undetected compromise of assets may occur as a result of:			
T1a	an authorized user performing actions the individual is not authorized to perform	X	X	X
T1b	an attacker (insider or outsider) masquerading as an authorized user and attempting to perform actions that individual is authorized to perform	X	X	X
T1c	an attacker (insider or outsider) gaining unauthorized access to information or resources by impersonating an authorized user.	X	X	X
T1d	an authorized or unauthorized user accidentally or intentionally blocking staff access to TOE devices	X	X	X
T1e	an unauthorized user gaining control of the TOE	X	X	X
T1f	an unauthorized user rendering the TOE inoperable	X	X	X
T1g	an unauthorized person attempting to bypass security			X
T1h	an unauthorized person repeatedly trying to guess identification and authentication data	X	X	
T1i	an unauthorized person using valid identification and authentication data fraudulently	X	X	X
T1j	an unauthorized person or external IT entity viewing, modifying, and/or deleting security relevant information transmitted to a remote authorized user or administrator			X
T2	an authorized user may access information or resources without having permission from the person who owns or is responsible for the information or resource	X	X	X
T3	An attacker may eavesdrop on or otherwise capture data being transmitted across a network:	X		
T3a	an unauthorized users performing traffic analysis	X		
T3b	an authorized or unauthorized user using residual information from previous information flows	X		
T4	An authorized user or unauthorized outsider consumes global resources in a way that compromises the ability of other authorized users to access or use those resources:	X		
T4a	circuit jamming (voice or data)	X		
T4b	DoS and DDoS attacks (voice or data)	X		
T4c	theft of service	X		
T5	A user may intentionally or accidentally transmit sensitive information to users who are not cleared to see it	X		
T6	A user may participate in the transfer of information either as originator or recipient and then subsequently deny having done so.	X		
T7	An authorized user may export information in soft- or hard-copy form, which the recipient subsequently handles in a manner that is inconsistent with its sensitivity designation.	X		
T8	The integrity and availability of information may be compromised due to:			
T8a	user errors, firmware errors, hardware errors, or transmission errors	X	X	X
T8b	the unauthorized modification or destruction of the information by an attacker	X	X	X
T8c	human errors or a failure of software, firmware, hardware or power supplies which causes an abrupt interruption to operations, resulting in the loss or corruption of critical data	X	X	X

Exhibit 14. Threat Assessment: Step 1 (continued)

Table x Potential Threats to Assets by TOE

#	Threat	TOE A	TOE B	TOE C
T8d	aging of storage media or improper storage or handling of storage media	X	X	X
T8e	an authorized user unwittingly introducing a virus into the system	X	X	X
T8f	an authorized user may introduce unauthorized software into the system	X	X	X
T8g	an authorized or unauthorized user inserting malicious code or backdoors	X	X	X
T8h	an unauthorized person reading, modifying, or destroying security critical configuration information	X	X	X
T8i	failure to perform adequate system backups	X	X	X
T8j	accidental or intentional deletion	X	X	X
T8k	insertion of bogus data	X	X	X
T8l	unauthorized modification of data (payload or header)	X	X	X
T9	An attacker could observe the legitimate use of a resource or service by a user, when the user wishes their use of that resource or service to be kept confidential	X		
T10	An authorized user may intentionally or accidentally observe stored information that the user is not cleared to see		X	X
T11	Security-critical components may be subject to physical attack and/or operational environmental failures, which may compromise security			X
T12	An authorized insider or unauthorized outsider may accidentally or intentionally cause security-relevant events not to be recorded or traceable:			
T12a	legitimate audit records lost or overwritten	X	X	X
T12b	audit records may not be attributed to time of occurrence	X	X	X
T12c	audit records may not be attributed to actual source of activity	X	X	X
T12d	people may not be held accountable for their actions because audit records are not reviewed	X	X	X
T12e	compromises of user or system resources may go undetected for long periods of time	X	X	X
T13	Weaknesses in the architecture, design, implementation, operation, or maintenance may precipitate security failures or compromises	X	X	X
T14	An authorized insider or unauthorized outsider may cause the improper restart and/or recovery from failure of hardware, software, or firmware that causes a security compromise	X	X	X
T15	Changes in operational environment may introduce or exacerbate vulnerabilities	X	X	X
T16	A knowledgeable adversary may circumvent unexpected limitations or latent defects in countermeasures and mitigation strategies	X	X	X
T17	The definition, implementation, and enforcement of access control rights and privileges may be done in a manner that undermines security		X	X
T18	Natural disasters or acts of war or terrorism could result in critical operations being interrupted or halted	X	X	X
T19	Compromise of assets may occur as a result of actions taken by careless, willfully negligent or hostile administrators or other privileged users:			

Exhibit 14. Threat Assessment: Step 1 (continued)

Table *x* Potential Threats to Assets by TOE

#	Threat	TOE A	TOE B	TOE C
T19a	improper operation of hardware, software, and/or firmware	X	X	X
T19b	premature hang-up of voice circuit	X		
T19c	premature shut-down of PVC or VPN	X	X	
T19d	OPSEC procedures are inadequate	X	X	X
T19e	OPSEC procedures poorly written	X	X	X
T19f	users and administrators unfamiliar with OPSEC procedures	X	X	X

2. *Accountability* — Accountability policies require the explicit association of individual entities (human and non-human) with specific actions. They include concepts such as identification, authentication, and auditing.
3. *Availability* — Availability policies mandate that a system must be available for use when needed. They require mechanisms to be in place to ensure that: (1) resources are available when requested, and (2) recovery mechanisms are effective when a failure occurs.
4. *Confidentiality* — Confidentiality polices prescribe the type and strength of encryption mechanisms to be used by asset sensitivity.
5. *Integrity* — Integrity polices focus on maintaining system and data integrity regardless of the system mode or state: start-up, shut-down, normal operations, preventive maintenance, emergency shut-down, degraded mode operations, and so forth.
6. *Secure Installation and Operation* — Secure Installation and Operation policies promote preventing compromise of IT resources through appropriate system documentation, regular training and security reviews, personnel security practices, and physical security practices.
7. *Transmission Protection* — Transmission Protection policies seek to protect information assets during transmission to and from external entities, whether trusted or untrusted.

Exhibit 16 presents sample OSPs.

3.5 Section 4: Security Objectives

The fourth required section of a PP, Security Objectives, provides a concise statement of the intended response to the Security Environment described in Section 3.[19] In other words, security objectives uphold all identified assumptions, counter all identified threats, and enforce all stated organizational security policies. The purpose of this section is to divide responsibilities between the TOE and the TOE environment. As the standard observes, the statement of Security Objectives serves to:[22]

Exhibit 15. Threat Assessment: Step 2

Table x Risk-Based Analysis of Potential Threats to Assets

#	Threat	Severity of Consequences (note 1)	Likelihood of Occurrence (note 2)	Risk Mitigation Priority
T1	An undetected compromise of assets may occur as a result of:			
T1a	an authorized user performing actions the individual is not authorized to perform	marginal to critical	occasional	high
T1b	an attacker (insider or outsider) masquerading as an authorized user and attempting to perform actions that individual is authorized to perform	marginal to critical	occasional	high
T1c	an attacker (insider or outsider) gaining unauthorized access to information or resources by impersonating an authorized user.	marginal to critical	occasional	high
T1d	an authorized or unauthorized user accidentally or intentionally blocking staff access to TOE devices	marginal to critical	occasional	high
T1e	an unauthorized user gaining control of the TOE	marginal to critical	remote	medium to high
T1f	an unauthorized user rendering the TOE inoperable	marginal to critical	remote	medium to high
T1g	an unauthorized person attempting to bypass security	Marginal to critical	frequent	medium to high
T1h	an unauthorized person repeatedly trying to guess identification and authentication data	marginal to critical	frequent	medium to high
T1i	an unauthorized person using valid identification and authentication data fraudulently	marginal to critical	probable	medium to high
T1j	an unauthorized person or external IT entity viewing, modifying, and/or deleting security relevant information transmitted to a remote authorized user or administrator	marginal to critical	occasional	medium to high
T2	An authorized user may access information or resources without having permission from the person who owns or is responsible for the information or resource	marginal to critical	remote	medium
T3	An attacker may eavesdrop on or otherwise capture data being transmitted across a network:			
T3a	unauthorized users performing traffic analysis	marginal	remote	low
T3b	an authorized or unauthorized user using residual information from previous information flows	marginal	remote	low
T4	An authorized user or unauthorized outsider consumes global resources in a way that compromises the ability of other authorized users to access or use those resources:			

Exhibit 15. Threat Assessment: Step 2 (continued)

Table *x* Risk-Based Analysis of Potential Threats to Assets

#	Threat	Severity of Consequences (note 1)	Likelihood of Occurrence (note 2)	Risk Mitigation Priority
T4a	circuit jamming (voice or data)	marginal to catastrophic	remote	high
T4b	DoS and DDoS attacks (voice or data)	marginal to catastrophic	remote	high
T4c	theft of service	marginal to catastrophic	remote	high
T5	A user may intentionally or accidentally transmit sensitive information to users who are not cleared to see it	marginal to critical	remote	medium
T6	A user may participate in the transfer of information either as originator or recipient and then subsequently deny having done so	marginal	remote	Low
T7	An authorized user may export information in soft- or hard-copy form which the recipient subsequently handles in a manner that is inconsistent with its sensitivity designation.	marginal to critical	occasional	high
T8	The integrity and availability of information may be compromised due to:			
T8a	user errors, firmware errors, hardware errors, or transmission errors	marginal to catastrophic	occasional	high
T8b	unauthorized modification or destruction of the information by an attacker	marginal to catastrophic	remote	medium
T8c	human errors or a failure of software, firmware, hardware or power supplies which causes an abrupt interruption to operations, resulting in the loss or corruption of critical data	marginal to catastrophic	remote	medium
T8d	aging of storage media or improper storage or handling of storage media	marginal to catastrophic	remote	medium
T8e	authorized user unwittingly introducing a virus into the system	marginal to catastrophic	frequent	high
T8f	authorized user introducing unauthorized software into the system	marginal to catastrophic	frequent	high
T8g	authorized or unauthorized user inserting malicious code or backdoors	marginal to catastrophic	occasional	medium
T8h	an unauthorized person reading, modifying, or destroying security critical configuration information	marginal to catastrophic	occasional	medium to high
T8i	failure to perform adequate system backups	marginal	occasional	medium
T8j	accidental or intentional deletion	marginal to critical	occasional	medium to high
T8k	insertion of bogus data	marginal to critical	occasional	medium to high
T8l	unauthorized modification of data (payload or header)	marginal to critical	occasional	medium to high

Exhibit 15. Threat Assessment: Step 2 (continued)

Table *x* Risk-Based Analysis of Potential Threats to Assets

#	Threat	Severity of Consequences (note 1)	Likelihood of Occurrence (note 2)	Risk Mitigation Priority
T9	An attacker could observe the legitimate use of a resource or service by a user, when the user wishes their use of that resource or service to be kept confidential	marginal to critical	occasional	high
T10	An authorized user may intentionally or accidentally observed store information that the user is not cleared to see	marginal to critical	occasional	medium
T11	Security-critical components may be subject to physical attack and/or operational environmental failures, which may compromise security	insignificant to catastrophic	improbable	low
T12	An authorized insider or unauthorized outsider may accidentally or intentionally cause security-relevant events not to be recorded or traceable:	marginal to catastrophic	remote	medium
T12a	legitimate audit records being lost or overwritten	marginal to catastrophic	remote	medium
T12b	audit records not being attributed to time of occurrence	marginal to catastrophic	remote	medium
T12c	audit records not being attributed to actual source of activity	marginal to catastrophic	remote	medium
T12d	people not held accountable for their actions because audit records are not reviewed	marginal to catastrophic	remote	medium
T12e	compromises of user or system resources going undetected for long periods of time	marginal to catastrophic	remote	medium
T13	Weaknesses in the architecture, design, implementation, operation, or maintenance may precipitate security failures or compromises	marginal to critical	remote	medium
T14	An authorized insider or unauthorized outsider may cause the improper restart and/or recovery from failure of hardware, software, or firmware that causes a security compromise	marginal to critical	remote	medium
T15	Changes in operational environment may introduce or exacerbate vulnerabilities	marginal to critical	remote	low
T16	A knowledgeable adversary may circumvent unexpected limitations or latent defects in countermeasures and mitigation strategies	marginal to critical	remote	medium
T17	The definition, implementation, and enforcement of access control rights and privileges may be done in a manner that undermines security	marginal to critical	remote	medium
T18	Natural disasters or acts of war or terrorism could result in critical operations being interrupted or halted	marginal to catastrophic	improbable	low

Exhibit 15. Threat Assessment: Step 2 (continued)

Table *x* Risk-Based Analysis of Potential Threats to Assets

#	Threat	Severity of Consequences (note 1)	Likelihood of Occurrence (note 2)	Risk Mitigation Priority
T19	Compromise of assets may occur as a result of actions taken by careless, willfully negligent or hostile administrators or other privileged users:			
T19a	improper operation of hardware, software, and/or firmware	marginal to catastrophic	remote	medium
T19b	premature hang-up of voice circuit	marginal to catastrophic	remote	medium
T19c	premature shut-down of PVC or VPN	marginal to catastrophic	remote	medium
T19d	OPSEC procedures are inadequate	marginal to catastrophic	remote	medium
T19e	OPSEC procedures poorly written	marginal to catastrophic	remote	medium
T19f	users and administrators unfamiliar with OPSEC procedures	marginal to catastrophic	remote	medium

Note 1: Standard severity definitions from IEC 61508 are used:

- *catastrophic*—loss of one or more major systems which may or may not be accompanied by fatalities and/or multiple severe injuries;
- *critical*—loss of a major system which may or may not be accompanied by a single fatality or severe injury;
- *marginal*—severe system damage which may or may not be accompanied by minor injuries;
- *insignificant*—system damage which may or may not be accompanied by single minor injury.

Note 2: Standard likelihood definitions from IEC 61508 are used:

- *frequent*—likely to occur frequently, 10^{-2};
- *probable*—will occur several times, 10^{-3};
- *occasional*—likely to occur several times over the life of a system, 10^{-4};
- *remote*—likely to occur at some time during the life of a system, 10^{-5};
- *improbable*—unlikely but possible to occur during the life of a system, 10^{-6};
- *incredible*—extremely unlikely to occur during the life of a system, 10^{-7}.

- Outline what the TOE will and will not do within the context of the TOE security environment.
- Scope the evaluation of the TOE.
- Drive the selection of security functional requirements and determination of the level of assurance needed.

Security Objectives are written for the TOE and the operational environment (IT and non-IT). Countermeasures deployed by the TOE satisfy TOE Security Objectives.[22] Technical measures implemented by the IT environment meet

Exhibit 16. Sample Organizational Security Policies

3.3 Organizational Security Policies

This subsection identifies the organizational security policies with which the target of evaluation (TOE) or the TOE environment must comply.

3.3.1 Access Control Policy

- P1 All data collected and produced by the TOE shall only be used for authorized purposes.
- P2 Authorized users and administrators of the TOE shall be eligible to access information that is collected, created, communicated, disseminated, processed, or stored by the TOE in accordance with their access control rights and privileges.
- P3 The system shall be capable of enforcing separation of duties through role-based access control that restricts users to specific data objects and to specific actions upon those objects.

3.3.2 Accountability Policy

- P4 Users of the TOE shall be accountable for their actions within the TOE.
- P5 User activity shall be monitored to the extent that sanctions can be applied when malfeasance occurs and to ensure that system controls are properly applied. All users will be notified that such monitoring may occur.

3.3.3 Availability Policy

- P6 The TOE shall be capable of providing resource allocation features having a measure of resistance to resource depletion.
- P7 The TOE shall provide fault tolerance, fail secure, and recovery features that provide a measure of survivability.

3.3.4 Confidentiality Policy

- P8 The confidentiality and privacy of user and system data shall be protected in accordance with its sensitivity and criticality.
- P9 User data shall be adequately marked to indicate the sensitivity of the information.

3.3.5 Integrity Policy

- P10 Data stored, generated, and processed by the TOE shall be protected from unauthorized modification, deletion, and insertion.
- P11 At start-up the TOE shall perform a self-check for the presence and correct operating capability of the TTSF and shall abort operations and generate an alarm in response to negative findings.
- P12 The TOE shall be capable of monitoring file integrity and generating alerts when file integrity is compromised.
- P13 The TOE shall be capable of removing or isolating malicious code and data from executable programs and communications traffic.

3.3.6 Secure Installation and Operation Policy

- P14 Analytical processes and information to derive conclusions about intrusions (imminent, known, or suspected) shall be applied and appropriate responses taken.
- P15 The TOE shall only be managed by authorized users.
- P16 The TOE shall be protected from unauthorized access to and disruption of TOE data and functions.
- P17 The TOE shall be able to interoperate with other IT systems with which it interfaces in a secure manner.
- P18 The TOE shall be managed such that its security functions are implemented and preserved throughout its operational lifetime.
- P19 The TOE must be physically protected.
- P20 Authorized users and system administrators shall be adequately trained.
- P21 Authorized users and system administrators must undergo appropriate background checks.
- P22 The TOE must be subjected to periodic security audits and vulnerability assessments.
- P23 The TOE shall have documentation describing security features, functions, and configuration parameters, and the residual risk associated with the use of the system.

3.3.7 Transmission Protection Policy

- P24 Data (user, network management, and security management) transmitted by the TOE shall be protected from unauthorized eavesdropping, modification, deletion, insertion, and replay.
- P25 Authorized users and administrators of the TOE shall not export information processed by the TOE without proper and explicit authorization.

Security Objectives for the IT environment,[22] while procedural measures (operational security, or OPSEC) achieve Security Objectives for the non-IT environment.[22] Security Objectives are implemented through a combination of SFRs; each SFR maps to one or more Security Objective while each Security Objective maps to at least one SFR.

Security Objectives are categorized as being preventive, detective, or corrective:[22]

- *Preventive objectives* — Objectives that prevent a threat from being carried out or limit the ways in which it can be carried out.
- *Detective objectives* — Objectives that detect and monitor the occurrence of events relevant to the secure operation of the TOE.
- *Corrective objectives* — Objectives that require the TOE to take action in response to potential security violations, anomalies, or other undesirable events, in order to preserve or return to a secure state or limit any damage.

This strategy corresponds to the chronology of threat control measures and the corresponding priorities established by the CC/CEM for preventing security vulnerabilities. The chronology of threat control measures consists of five phases (see Exhibit 17):

1. *Anticipate/Prevent* — Threat types and sources are anticipated *a priori* so that proactive preventive action can be taken to reduce the likelihood of a threat being instantiated and the severity of its consequences.
2. *Detect* — The TOE detects all imminent known or suspected attacks, whether or not they are successful.
3. *Characterize* — Attacks are characterized so that appropriate short-term responses and long-term recovery actions can be formulated.
4. *Respond/Contain Consequences* — Short-term responses are implemented to quickly isolate and contain the consequences of threat instantiation.
5. *Recover* — Long-term recovery measures ("lessons learned") are deployed to eliminate or mitigate the consequences of the same or similar threats in the future.

The chronology of threat control measures parallels the priorities the CC/CEM has established for preventing security vulnerabilities (see Exhibit 18):

1. *Elimination* — (a) Exposing security vulnerabilities through ongoing security assurance activities; (b) removing security vulnerabilities by the (re)specification of correct, complete, consistent, unambiguous, and verifiable SFRs and SARs; and (c) neutralizing security vulnerabilities by the (re)development of a resilient security architecture.
2. *Minimization* — Reducing the likelihood of vulnerabilities being exploited and the severity of the consequences from threat instantiation by designing and deploying a robust defense and in-depth security architecture.

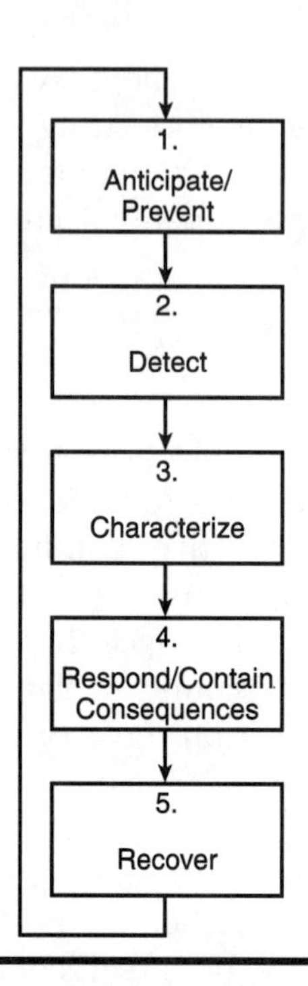

Exhibit 17. Chronology of Threat Control Measures

3. *Monitoring* — Detecting any attempt to exploit residual vulnerabilities through ongoing diverse multisource monitoring, reporting, and analysis of anomalous security events and rapid preplanned informed responses to contain consequences.
4. *Communication* — Full disclosure to end users, system administrators, and system owners about residual security vulnerabilities. This is accomplished by (a) in-depth training about the correct operation and use of a system and its security features, and (b) comprehensive warnings about residual security vulnerabilities and consequences of misuse.

Exhibit 19 presents sample security objectives for a TOE, while Exhibit 20 presents sample Security Objectives for the environment. In the case of a composite TOE, Security Objectives are assigned to the appropriate component TOE. Likewise, Security Objectives can be assigned to functional packages for a monolithic TOE.

3.6 Section 5: Security Requirements

The fifth required section of a PP is Security Requirements. Broadly speaking, these requirements specify the presence of desired behavior and conversely the absence of undesired behavior.[19] Security requirements are described in four subsections: Security Functional Requirements (SFRs), Security Assurance Requirements (SARs), Security Requirements for the IT Environment, and Security Requirements for the Non-IT Environment.

Exhibit 18. Priorities for Preventing Security Vulnerabilities

1st Elimination
Expose, remove, and neutralize security vulnerabilities.

2nd Minimization
Reduce the likelihood of vulnerabilities being exploited and the severity of the consequences from threat instantiation.

3rd Monitoring
Detect any attempt to exploit residual vulnerabilities.

4th Communication
Provide full disclosure to end users, system administrators, and system owners about residual security vulnerabilities.

Exhibit 19. Sample Security Objectives for TOE

Table x Security Objectives by TOE

#	Objective	Type*	TOE A	TOE B	TOE C
O1	The TSF must ensure that only authorized users gain access to the TOE and its resources by enforcing discretionary access controls.	P	X	X	X
O2	The TSF must ensure that any information contained in a protected resource is not released when the resource is recycled.	P		X	X
O3	The TSF will record all security relevant events and generate alarms when necessary.	D	X	X	X
O4	The TSF will protect the confidentiality of information when it is stored (online or archive), processed, and transmitted.	P	X	X	X
O5	The TSF will detect the loss of system or data integrity.	D	X	X	X
O6	Data exported will have sensitivity labels that are an accurate representation of the corresponding internal sensitivity labels.	P		X	
O7	The TOE will return to a known secure state following a system fault, failure or compromise.	C	X	X	X
O8	The TSF will isolate any network segment or system resource experiencing an attack or virus or worm infection.	C	X	X	X
O9	The TOE will protect itself against external interference, tampering, and attempts to bypass security functions.	P	X	X	X
O10	The TSF will control the consumption of global resources, including the number of concurrent sessions.	P	X	X	
O11	The TSF will prevent the TOE from becoming a vehicle for attacking other systems.	P	X		
O12	The TOE will not be used to decrease the availability of other systems.	P	X	X	X

* P—preventive security objective

D—detective security objective

C—corrective security objective

Exhibit 20. Sample Security Objectives for the Environment

Table y Security Objectives for the Operational Environment

#	Objective	Type*	IT Environment	Non-IT Environment
O13	System activity audit records will be reviewed daily and stored online for 7 days, offline for 90 days.	D		X
O14	The system security administrator will implement access control rights and privileges as directed.	P		X
O15	The TOE environment will support the enterprise-wide cryptographic infrastructure.	P	X	
O16	Internal and external TOE entities shall be deployed to monitor for and provide protection against natural and manmade environmental threats (fire, flood, humidity, dust, vibration, earthquakes, temperature fluctuations, power fluctuations, etc.)	P, D, C	X	X
O17	Current and complete documentation and training will be provided to end users and system administrators on a regular basis.	P		X
O18	The TOE will be protected from malicious physical attacks, tampering, unauthorized modification, destruction, and theft.	P, D	X	X
O19	The TOE will be connected to a reliable time source to allow proper synchronization of resources.	P	X	
O20	The TOE will be delivered, installed, managed, and operated in a manner that maintains the security posture.	P	X	X

* P—preventive security objective

D—detective security objective

C—corrective security objective.

The introduction to Section 5 of a PP contains a mandatory statement, which can take only three valid forms:[22]

> *Option 1* — All of the requirements in this PP apply to the TOE itself, as opposed to the TOE environment.
> *Option 2* — The IT security requirements section provides detailed security requirements, in separate subsections, for the TOE and the IT environment.
> *Option 3* — The security requirements section provides detailed security requirements, in separate subsections, for the TOE, the IT environment, and the non-IT environment.

Option 1 is used if the security requirements in Section 5 of the PP only apply to the TOE; no security requirements exist for the IT or non-IT environment. In this instance, a PP would not contain Subsections 5.3 or 5.4. Option 2 is used if the security requirements in Section 5 of the PP only apply to the TOE and the IT environment. As a result, a PP would not contain Subsection

5.4. Option 3 is used if the security requirements in Section 5 of the PP apply to the TOE, the IT environment, and the non-IT environment. In other words, Section 5 of a PP would contain all four subsections.

If an EAL greater than EAL 2 is specified or if AVA_SOF.1 is included in the security assurance requirements, the following clause is added at the end of the mandatory sentences cited above:[22]

> ...including strength of function requirements for TOE security functions realized by a probabilistic or permutational mechanisms.

3.6.1 Security Functional Requirements (SFRs)

Security functional requirements (SFRs) implement the security objectives stated in Section 4 of a PP. As stated previously, each SFR maps to one or more security objectives, while each security objective maps to at least one SFR. The selection of SFRs, or the type of protection needed, is influenced by three key factors:

1. Sensitivity and value of the assets being protected
2. Criticality of the mission the system performs
3. Consequences from the assets or system being lost, compromised, misappropriated, corrupted, destroyed, misused, or rendered inoperable or unavailable for an extended period of time

Secondary factors that may also affect the selection of SFRs include cost and schedule constraints. The goal is to select SFRs that meet all stated security objectives without over- or under-protecting a system and its assets. A common practice is to include a table at the beginning of this section that maps functional components to (1) TOEs for a composite TOE, or (2) FPs for a component TOE.

Exhibit 21 depicts the decision-making process for selecting SFRs to include in a PP. The selection of SFRs should not be a cursory or haphazard exercise ("That looks good ... I think we need some of that"). Rather, a systematic and methodical decision-making process is followed. The first step is to select a security objective from those stated in Section 4 of the PP. Then, a determination is made about the type and purpose of the security objective. For instance, is the objective for the TOE, the IT environment, or the non-IT environment? Is the objective preventive, detective, or corrective?

The next step is to identify the corresponding security functional class for this objective. To illustrate, if the objective mentions "auditing," then the FAU functional class is picked. The appropriate functional family within this class is then ascertained. This involves determining whether the functional family conforms to a preventive, detective, or corrective requirement. Exhibit 22 maps functional families to security objectives by categorizing them as being preventive, detective, or corrective. Detective requirements are further delineated as detecting or characterizing attacks. Corrective requirements are further

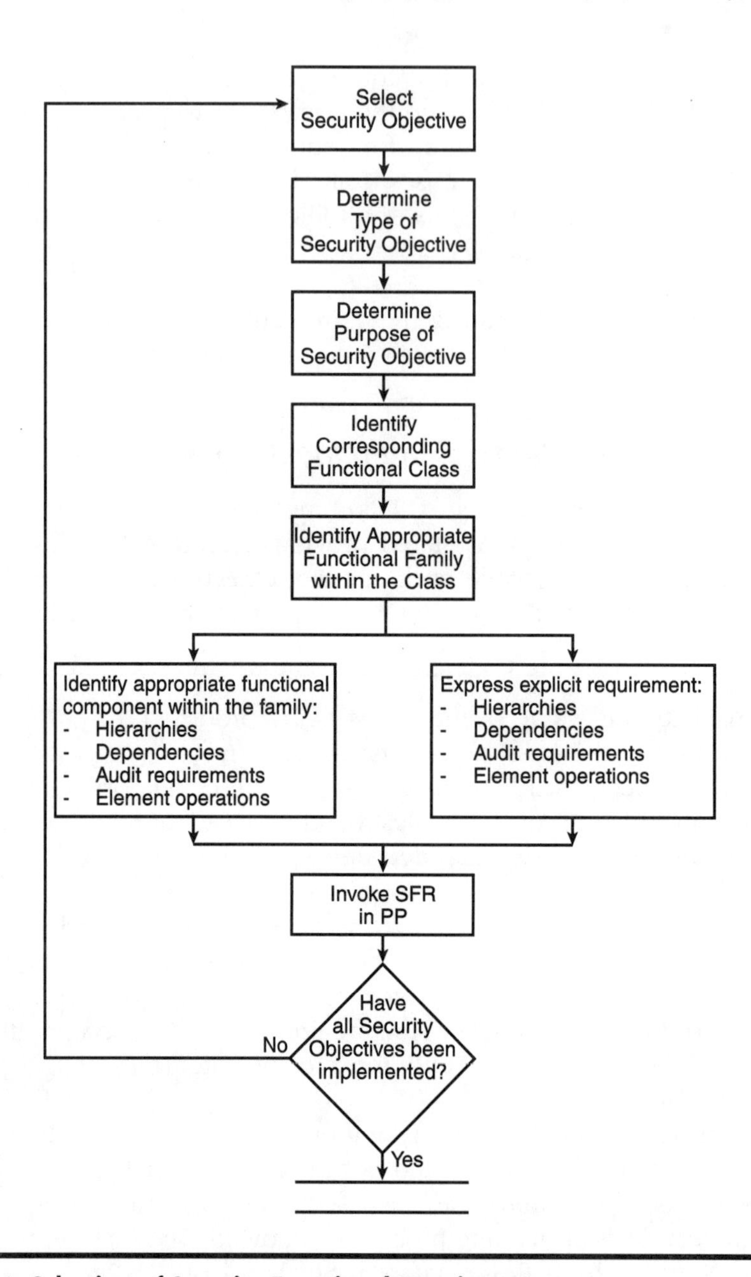

Exhibit 21. Selection of Security Functional Requirements

delineated as supporting a short-term response or long-term recovery measure, consistent with the threat control chronology. Note that some families fall into more than one category.

Continuing through the decision-making process, the appropriate functional component to use, within the family identified, is determined next. If an appropriate family or component cannot be identified, the PP author may specify an explicit requirement. When the CC were originally released, some confusion arose regarding whether or not the SFRs and SARs stated in the

Exhibit 22. Security Functional Requirements (SFRs) Mapped to Security Objectives

SFR Class	SFR Family	Security Objective				
		Prevent	Detect		Correct	
		Anticipate, Prevent	Detect	Characterize	Respond, Contain Consequences	Recover
FAU	ARP				X	
	GEN		X			
	SAA			X		
	SAR			X		
	SEL		X			
	STG		X			
FCO	NRO	X				
	NRR	X				
FCS	CKM	X				
	COP	X				
FDP	ACC	X				
	ACF	X				
	DAU	X				
	ETC	X				
	IFC	X				
	IFF	X				
	ITC	X				
	ITT	X	X		X	
	RIP	X				
	ROL				X	X
	SDI		X	X	X	
	UCT	X				
	UIT	X	X	X		X
FIA	AFL		X	X	X	
	ATD	X				
	SOS	X				
	UAU	X	X			
	UID	X				
	USB	X				
FMT	MOF	X				
	MSA	X				
	MTD	X				
	REV	X			X	
	SAE	X				

Exhibit 22. Security Functional Requirements (SFRs) Mapped to Security Objectives (continued)

		Security Objective				
		Prevent	Detect		Correct	
SFR Class	SFR Family	Anticipate, Prevent	Detect	Characterize	Respond, Contain Consequences	Recover
	SMF*	X				
	SMR	X				
FPR	ANO	X				
	PSE	X				
	UNL	X				
	UNO	X				
FPT	AMT	X				
	FLS	X				X
	ITA	X				
	ITC	X				
	ITI		X	X	X	
	ITT	X	X	X	X	
	PHP		X	X	X	
	RCV					X
	RPL		X	X	X	
	RVM	X				
	SEP	X				
	SSP	X				
	STM			X		
	TDC	X				
	TRC	X				
	TST	X				
FRU	FLT				X	X
	PRS	X				
	RSA	X				
FTA	LSA	X				
	MCS	X				
	SSL	X				
	TAB	X				
	TAH	X				
	TSE	X			X	X
FTP	ITC	X				
	TRP	X				

* Per Final Interpretation 095.

standard were the only options from which a PP author could choose. Given the incredibly rapid advancements in IT and the extreme diversity of IT systems and security needs, it would be impossible for any standard to include all the requisite SFRs and SARs. The CC Implementation Management Board (CCIMB) recognized this situation and made a provision for a PP author to include SFRs and SARs that are not contained in Part 2 or Part 3 of the standard. This feature is known as explicit requirements. The use of explicit SFRs is noted in the PP Identification field (1.1.6: Common Criteria Conformance Claim and Version) as "Part 2 extended". As the standard states:[20]

> ISO/IEC 15408 and the associated functional security requirements described herein are not meant to be a definitive answer to all the problems of IT security. Rather, the standard offers a set of well-understood functional security requirements that can be used to create trusted products or systems. …This part of ISO/IEC 15408 does not presume to include all possible functional security requirements but rather contains those that are known and agreed to be of value … at the time of release. Since the understanding and needs of consumers may change, the functional requirements in this Part will need to be maintained … PP/ST authors may choose to consider using functional requirements not taken from the standard.

Certain criteria are levied on the construction and use of explicit requirements by the assurance class APE_SRE, Protection Profile Evaluation — Explicitly Stated IT Security Requirements. Specifically, explicit requirements must be:[21]

- Clearly identified as an explicit requirement
- Unambiguously expressed
- At the appropriate level of detail
- Self-contained (no external dependencies)
- Measurable
- Verifiable

In addition the use of explicit requirements must be justified. An explanation of why the PP author must state an explicit requirement, rather than use a standard SFR, is included in Section 7 of the PP (Rationale). This reinforces the need to be certain that a standard SFR does not already exist that will meet the PP author's needs; explicit requirements should not be created simply because the PP author is not familiar with current CC SFRs.

The development of explicit requirements adheres to the conventions used to express standard functional classes, families, components, and elements. For every explicit functional requirement, a corresponding explicit assurance requirement must be developed to explain how the fulfillment of that requirement will be evaluated: Developer actions, content and presentation criteria, and evaluator actions must be defined. The example below illustrates the specification of explicit functional requirements for two scenarios: (1) adding a new family to an existing class, and (2) adding a new class and family.

Example 1: adding a new family to an existing class

Explicit Requirement

FPT_ISO.1, Isolation of IP Services

FPT_ISO.1.1	*The TSF shall disconnect or isolate any authorized or unauthorized IP service interconnection, network segment, or digital demarcation point within 10 minutes when directed to do so by a designated Government authority.*

Example 2: adding a new class and family

Explicit requirement

FEX Extranet Services

FEX_STP.1 Extranet Services to Trusted Partners

FEX_STP.1.1	*The TSF shall provide extranet services between identified locations with trusted partners.*
FEX_STP.1.2	*The TSF shall provide enhanced extranet IP filtering and routing services between identified locations and trusted partners.*
FEX_STP.1.3	*The TSF shall provide extranet services that isolate incoming and outgoing extranet traffic within the appropriate extranet segment.*

If an SFR relies upon statistical or quantitative mechanisms to perform identification, authentication, integrity, or encryption functions, a minimum strength of function (SOF) must be specified for that capability. An SOF can be specified at the beginning of Section 5 of a PP or through a refinement operation, which is discussed below. If the SOF is specified at the beginning of Section 5, the same level applies to all security functions. Expressing the SOF as part of a refinement operation allows the PP author to state a different SOF for each function. As the standard states:[19]

> Where AVA_SOF.1 is included in the TOE security assurance requirements (e.g. EAL 2 or higher), the statement of TOE security functional requirements shall include a minimum strength level for the TOE security functions realized by a probabilistic or permutational mechanism (e.g. password or hash function). All such functions shall meet this minimum level. The level shall be one of the following: SOF-basic, SOF-medium, or SOF-high. The selection of the level shall be consistent with the identified security objectives for the TOE.

The first five steps in the selection process shown in Exhibit 21 are fairly straightforward, once the user becomes familiar with the catalog of SFRs contained in ISO/IEC 15408-2(12-1999). The sixth step, however, requires more attention to detail and greater fluency in the CC syntax and notation. At this point, four central characteristics of the candidate components are analyzed, specifically:

1. Hierarchies
2. Dependencies
3. Audit requirements
4. Operations

Hierarchical relationships, if any, between components of a family are explained in Part 2 of the CC standard. Hierarchies represent increasing strength of capability of security requirements that share a common purpose. The customer selects the appropriate strength needed based on stated security objectives and the threats they counter. Following each component mnemonic and name, Part 2 of the CC standard contains an entry that reads "Hierarchical to". If the component is a more robust implementation of a preceding component, the lower component is listed. If not, the standard states "Hierarchical to: No other components". Lateral relationships between components are depicted in a diagram under the component leveling description of each family. Exhibit 23 illustrates these concepts using the FAU_SAA (security audit analysis) functional family. FAU_SAA.1, FAU_SAA.3, and FAU_SAA.4 progressively increase the strength of a security audit analysis capability. FAU_SAA.2 provides a lateral anomaly detection capability. In this example, a customer would select one component from the FAU_SAA.1, FAU_SAA.3, FAU_SAA.4 chain or one component from the FAU_SAA.1, FAU_SAA.2 chain. Six choices are valid:

Option 1 — FAU_SAA.1 for a basic audit monitoring capability
Option 2 — FAU_SAA.2 for a medium anomaly detection capability
Option 3 — FAU_SAA.3 for a medium attack heuristics capability
Option 4 — FAU_SAA.3 with FAU_SAA.2 for medium attack heuristics and anomaly detection capabilities
Option 5 — FAU_SAA.4 for an advanced attack heuristics capability
Option 6 — FAU_SAA.4 with FAU_SAA.2 for an advanced attack heuristics capability and a medium anomaly detection capability

After hierarchy issues have been evaluated, dependencies are resolved. An SFR is classified as being a principal SFR or a supporting SFR. A principal SFR is an SFR that directly satisfies the security objectives of a TOE.[22] In contrast, a supporting SFR is an SFR that does not directly satisfy the security objectives of a TOE but rather provides support to a principal SFR, thereby indirectly helping to satisfy TOE security objectives.[22] The relationship between a principal SFR and a supporting SFR is referred to as a dependency. In a few instances, a principal SFR has a dependency on a supporting SAR as well. Dependencies can be internal or external to the family or class of

I. Component Leveling Example

FAU_SAA Security audit analysis

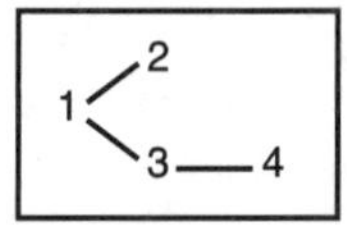

Interpretation:
- Component 1, FAU_SAA.1, is the lowest level component.
- Component 2, FAU_SAA.2, is higher than component 1 and lateral to component 3.
- Component 3, FAU_SAA.3, is higher than component 1 and lateral to component 2.
- Component 4, FAU_SAA.4, is higher than components 1 and 3.

II. Hierarchical to Example

FAU_SAA.1 Potential violation analysis
Hierarchical to: no other components

FAU_SAA.2 Profile based anomaly detection
Hierarchical to: FAU_SAA.1

FAU_SAA.3 Simple attack heuristics
Hierarchical to: FAU_SAA.1

FAU_SAA.4 Complex attack heuristics
Hierarchical to: FAU_SAA.3

Exhibit 23. Functional Hierarchy Example[20]

the principal SFR. Dependencies arise when a component is not entirely self-sufficient;[19] that is, in order to fully implement a capability, an SFR requires the underlying feature of another SFR or SAR. A dependency may take several forms:

- *Direct* — Primary dependency on a supporting SFR or SAR
- *Indirect* — Secondary dependency that occurs as a result of resolving a dependency of a supporting SFR or SAR
- *Reciprocal* — Bidirectional dependency between two principal SFRs
- *Multiple* — Principal SFR that is dependent upon more than one supporting SFRs or SARs
- *Multiplayer* —Chain of direct and indirect dependencies for a single principal SFR
- *Optional* — Direct dependency in which the customer chooses from a set of supporting SFRs or SARs to resolve the dependency

Exhibit 24 lists all functional dependencies by class and component. Dependencies are categorized as internal or external and as direct or indirect. Optional dependencies are shown, as well.

Some confusion was expressed when the CC were first issued regarding whether or not the list of dependencies cited in the standard were to be copied into a PP verbatim; that is not the case. Rather, the dependencies are to be resolved, and the SFRs and SARs selected to resolve them are to be included in the appropriate part of Section 5 of the PP, like any other requirement. Dependencies must be resolved unless an adequate justification for not doing so is provided in Section 7 of the PP (Rationale).

Exhibit 24. Functional Dependencies[20]

Class/ Component	None	Internal Dependencies		Dependencies with Options*	External Dependencies	
		Direct	Indirect		Direct	Indirect
FAU_ARP.1	—	FAU_SAA.1	FAU_GEN.1	—	—	FPT_STM.1
FAU_GEN.1	—	—	—	—	FPT_STM.1	—
FAU_GEN.2	—	FAU.GEN.1	—	—	FIA_UID.1	FPT_STM.1
FAU_SAA.1	—	FAU.GEN.1	—	—	—	FPT_STM.1
FAU_SAA.2	—	—	—	—	FIA_UID.1	—
FAU_SAA.3 FAU_SAA.4	X	—	—	—	—	—
FAU_SAR.1	—	FAU.GEN.1	—	—	—	FPT_STM.1
FAU_SAR.2 FAU_SAR.3	—	FAU_SAR.1	FAU_GEN.1	—	—	FPT_STM.1
FAU_SEL.1	—	FAU_GEN.1	—	—	FMT_MTD.1	FIA_UID.1 FMT_SMR.1 FPT_STM.1
FAU_STG.1 FAU_STG.2	—	FAU_GEN.1	—	—	—	FPT_STM.1
FAU_STG.3 FAU_STG.4	—	FAU_STG.1	FAU_GEN.1	—	—	FPT_STM.1
FCO_NRO.1 FCO_NRO.2	—	—	—	—	FIA_UID.1	—
FCO_NRR.1 FCO_NRR.2	—	—	—	—	FIA_UID.1	—
FCS_CKM.1	—	FCS_CKM.4	—	FCS_CKM.2 FCS_COP.1	FMT_MSA.2	**ADV_SPM.1** FDP_ACC.1 FDP_ACF.1 FDP_IFC.1 FDP_IFF.1 FDP_ITC.1 FIA_UID.1 FMT_MSA.1 FMT_MSA.3 FMT_SMR.1
FCS_CKM.2 FCS_CKM.3 FCS_COP.1	—	FCS_CKM.4		FDP_ITC.1 FCS_CKM.1	FMT_MSA.2	**ADV_SPM.1** FDP_ACC.1 FDP_ACF.1 FDP_IFC.1 FDP_IFF.1 FIA_UID.1 FMT_MSA.1 FMT_MSA.3 FMT_SMR.1
FCS_CKM.4	—	—	FCS_CKM.2 FCS_COP.1	FDP_ITC.1 FCS_CKM.1	FMT_MSA.2	**ADV_SPM.1** FDP_ACC.1 FDP_ACF.1 FDP_IFC.1 FDP_IFF.1 FIA_UID.1 FMT_MSA.1 FMT_MSA.3 FMT_SMR.1
FDP_ACC.1 FDP_ACC.2	—	FDP_ACF.1	FDP_ACC.1 FDP_IFC.1 FDP_IFF.1	—	—	FIA_UID.1 FMT_MSA.1 FMT_MSA.3 FMT_SMR.1

Exhibit 24. Functional Dependencies[20] (continued)

Class/ Component	None	Internal Dependencies		Dependencies with Options*	External Dependencies	
		Direct	Indirect		Direct	Indirect
FDP_ACF.1	—	FDP_ACC.1	FDP_IFC.1 FDP_IFF.1	FMT_MSA.3	—	FIA_UID.1 FMT_MSA.1 FMT_SMR.1
FDP_DAU.1	X	—	—	—	—	—
FDP_DAU.2	—	—	—	FIA_UID.1	—	—
FDP_ETC.1 FDP_ETC.2	—	—	FDP_ACF.1 FDP_IFF.1	FDP_ACC.1 FDP_IFC.1	—	FIA_UID.1 FMT_MSA.1 FMT_MSA.3 FMT_SMR.1
FDP_IFC.1 FDP_IFC.2	—	FDP_IFF.1	FDP_ACC.1 FDP_ACF.1 FDP_IFC.1	—	—	FIA_UID.1 FMT_MSA.1 FMT_MSA.3 FMT_SMR.1
FDP_IFF.1 FDP_IFF.2	—	FDP_IFC.1	FDP_ACC.1 FDP_ACF.1 FDP_IFF.1	—	FMT_MSA.3	FIA_UID.1 FMT_MSA.1 FMT_SMR.1
FDP_IFF.3 FDP_IFF.4 FDP_IFF.6	—	FDP_IFC.1	FDP_ACC.1 FDP_ACF.1 FDP_IFF.1	—	**AVA_CCA.1**	**ADV_FSP.2** **ADV_IMP.2** **AGD_ADM.1** **AGD_USR.1** FIA_UID.1 FMT_MSA.1 FMT_MSA.3 FMT_SMR.1
FDP_IFF.5	—	FDP_IFC.1	FDP_ACC.1 FDP_ACF.1 FDP_IFF.1	—	**AVA_CCA.3**	**ADV_FSP.2** **ADV_IMP.2** **AGD_ADM.1** **AGD_USR.1** FIA_UID.1 FMT_MSA.1 FMT_MSA.3 FMT_SMR.1
FDP_ITC.1	—	—	FDP—ACF.1 FDP_IFF.1	FDP_ACC.1 FDP_IFC.1	FMT_MSA.3	FIA_UID.1 FMT_MSA.1 FMT_SMR.1
FDP_ITC.2	—	—	FDP—ACF.1 FDP_IFF.1	FDP_ACC.1 FDP_IFC.1 FTP_ITC.1 FTP_TRP.1	FPT_TDC.1	FIA_UID.1 FMT_MSA.1 FMT_MSA.3 FMT_SMR.1
FTP_ITT.1 FTP_ITT.2	—	—	FDP—ACF.1 FDP_IFF.1	FDP_ACC.1 FDP_IFC.1	—	FIA_UID.1 FMT_MSA.1 FMT_MSA.3 FMT_SMR.1
FTP_ITT.3	—	FDP_ITT.1	FDP—ACF.1 FDP_IFF.1	FDP_ACC.1 FDP_IFC.1	—	FIA_UID.1 FMT_MSA.1 FMT_MSA.3 FMT_SMR.1
FTP_ITT.4	—	FDP_ITT.2	FDP—ACF.1 FDP_IFF.1	FDP_ACC.1 FDP_IFC.1	—	FIA_UID.1 FMT_MSA.1 FMT_MSA.3 FMT_SMR.1
FDP_RIP.1 FDP_RIP.2	X	—	—	—	—	—

Exhibit 24. Functional Dependencies[20] (continued)

Class/ Component	None	Internal Dependencies		Dependencies with Options*	External Dependencies	
		Direct	Indirect		Direct	Indirect
FDP_ROL.1 FDP_ROL.2	—	—	FDP—ACF.1 FDP_IFF.1	FDP_ACC.1 FDP_IFC.1	—	FIA_UID.1 FMT_MSA.1 FMT_MSA.3 FMT_SMR.1
FDP_SDI.1 FDP_SDI.2	X	—	—	—	—	—
FDP_UCT.1	—	—	FDP—ACF.1 FDP_IFF.1	FTP_ITC.1 FTP_TRP.1 FTP_ACC.1 FTP IFC.1	—	FIA_UID.1 FMT_MSA.1 FMT_MSA.3 FMT_SMR.1
FDP_UIT.1	—	—	FDP—ACF.1 FDP_IFF.1	FDP_ACC.1 FDP_IFC.1 FTP_ITC.1 FTP TRP.1	—	FIA_UID.1 FMT_MSA.1 FMT_MSA.3 FMT_SMR.1
FDP_UIT.2 FDP_UIT.3	—	FDP_UIT.1	FDP—ACF.1 FDP_IFF.1	FDP_ACC.1 FDP_IFC.1	FTP_ITC.1	FIA_UID.1 FMT_MSA.1 FMT_MSA.3 FMT_SMR.1
FIA_AFL.1	—	FIA_UAU.1	—	—	—	FIA_UID.1
FIA_ATD.1	X	—	—	—	—	—
FIA_SOS.1 FIA_SOS.2	X	—	—	—	—	—
FIA_UAU.1 FIA_UAU.2	—	FIA_UID.1	—	—	—	—
FIA_UAU.3 FIA_UAU.4 FIA_UAU.5 FIA_UAU.6	X	—	—	—	—	—
FIA_UAU.7	—	FIA_UAU.1	—	—	—	FIA_UID.1
FIA_UID.1 FIA_UID.2	X	—	—	—	—	—
FIA_USB.1	—	FIA_ATD.1	—	—	—	—
FMT_MOF.1	—	FMT_SMF.1 ** FMT_SMR.1	—	—	—	FIA_UID.1
FMT_MSA.1	—	FMT_SMF.1 ** FMT_SMR.1	FMT_MSA.3	FDP_ACC.1 FDP_IFC.1	—	FDP_ACF.1 FDP_IFF.1 FIA_UID.1
FMT_MSA.2	—	FMT_MSA.1 FMT_SMR.1	FMT_MSA.3	FDP_ACC.1 FDP_IFC.1	**ADV_SPM.1**	ADV_FSP.1 FDP_ACF.1 FDP_IFF.1 FIA_UID.1
FMT_MSA.3	—	FMT_MSA.1 FMT_SMR.1	—	—	—	FDP_ACC.1 FDP_ACF.1 FDP_IFC.1 FDP_IFF.1 FIA_UID.1
FMT_MTD.1	—	FMT_SMF.1 ** FMT_SMR.1	—	—	—	FIA_UID.1
FMT_MTD.2	—	FMT_MTD.1 FMT_SMR.1	—	—	—	FIA_UID.1

Exhibit 24. Functional Dependencies[20] (continued)

Class/ Component	None	Internal Dependencies		Dependencies with Options*	External Dependencies	
		Direct	Indirect		Direct	Indirect
FMT_MTD.3	—	FMT_MTD.1	—	—	**ADV_SPM.1**	**ADV_FSP.1** FIA_UID.1 FMT_SMR.1
FMT_REV.1	—	FMT_SMR.1	—	—	—	FIA_UID.1
FMT_SAE.1	—	FMT_SMR.1	—	—	FPT_STM.1	FIA_UID.1
FMT_SMF.1 **	X	—	—	—	—	—
FMT_SMR.1 FMT_SMR.2	—	—	—	—	FIA_UID.1	—
FMT_SMR.3	—	FMT_SMR.1	—	—	—	FIA_UID.1
FPT_ANO.1 FPT_ANO.2	X	—	—	—	—	—
FPR_PSE.1 FPR_PSE.3	X	—	—	—	—	—
FPR_PSE.2	—	—	—	—	FIA_UID.1	—
FPR_UNL.1	X	—	—	—	—	—
FPR_UNO.1 FPR_UNO.2 FPR_UNO.4	X	—	—	—	—	—
FPR_UNO.3	—	FPR_UNO.1	—	—	—	—
FPT_AMT.1	X	—	—	—	—	—
FPT_FLS.1	—	—	—	—	**ADV_SPM.1**	**ADV_FSP.1**
FPT_ITA.1	X	—	—	—	—	—
FPT_ITC.1	X	—	—	—	—	—
FPT_ITI.1 FPT_ITI.2	X	—	—	—	—	—
FPT_ITT.1 FPT_ITT.2	X	—	—	—	—	—
FPT_ITT.3	—	FPT_ITT.1	—	—	—	—
FPT_PHP.1 FPT_PHP.2		—	—	—	FMT_MOF.1	FIA_UID.1 FMT_SMR.1
FPT_PHP.3	X	—	—	—	—	—
FPT_RCV.1 FPT_RCV.2 FPT_RCV.3	—	FPT_TST.1	FPT_AMT.1	—	**AGD_ADM.1** **ADV_SPM.1**	**ADV_FSP.1**
FPT_RCV.4	—	—	—	—	**ADV_SPM.1**	**ADV_FSP.1**
FPT_RPL.1	X	—	—	—	—	—
FPT_RVM.1	X	—	—	—	—	—
FPT_SEP.1 FPT_SEP.2 FPT_SEP.3	X	—	—	—	—	—
FPT_SSP.1 FPT_SSP.2	—	FPT_ITT.1	—	—	—	—
FPT_STM.1	X	—	—	—	—	—
FPT_TDC.1	X	—	—	—	—	—
FPT_TRC.1	—	FPT_ITT.1	—	—	—	—
FPT_TST.1	—	FPT_AMT.1	—	—	—	—

Exhibit 24. Functional Dependencies[20] (continued)

Class/ Component	None	Internal Dependencies		Dependencies with Options*	External Dependencies	
		Direct	Indirect		Direct	Indirect
FRU_FLT.1 FRU_FLT.2	—	—	—	—	FPT_FLS.1	—
FRU_PRS.1 FRU_PRS.2	X	—	—	—	—	—
FRU_RSA.1 ** FRU_RSA.2	X	—	—	—	—	—
FTA_LSA.1	X	—	—	—	—	—
FTA_MCS.1 FTA_MCS.2	X	—	—	—	FIA_UID.1	—
FTA_SSL.1 FTA_SSL.2	—	—	—	—	FIA_UAU.1	FIA_UID.1
FTA_SSL.3	X	—	—	—	—	—
FTA_TAB.1	X	—	—	—	—	—
FTA_TAH.1	X	—	—	—	—	—
FTA_TSE.1	X	—	—	—	—	—
FTP_ITC.1	X	—	—	—	—	—
FPT_TRP.1	X	—	—	—	—	—

Note: **bold** indicates a dependency with a security assurance requirement.

* One component within the cell must be chosen.
**Per Final Interpretation 065.
X no dependencies
– not applicable

Audit requirements are developed from the list of potential auditable events identified in ISO/IEC 15408–2 in the description of each component. Part 2 of the CC states up to four hierarchical options for capturing auditable events: minimal, basic, detailed, and not specified. *Not specified* means that the PP author, not the CC standard, specifies the discrete events to be audited. Note that minimal, basic, and detailed audit requirements are incremental; each higher level of audit includes the levels below it. Three considerations should be taken into account when assigning auditable events:[22]

1. Is the benefit of collecting the information worth the impact on performance?
2. If the information is collected, will the administrator have sufficient resources (e.g., time and tool support) to effectively analyze the data?
3. What are the likely costs of managing or archiving the data collected?

The example that follows illustrates: (1) how to capture audit requirements in a PP, and (2) the relationship between auditable events listed for a functional component and the selection and assignment operations for the FAU_GEN.1 component. Assume the PP author wants to use the FPT_RCV.2 component.

ISO/IEC 15408–2 contains the following description of potential auditable events for FPT_RCV:[20]
[TS: refer to manuscript for general appearance of this section.]

Audit: FPT_RCV.1, FPT_RCV.2, FPT_RCV.3

The following actions should be auditable if FAU_GEN Security audit data generation is included in the PP/ST:

- *Minimal:* The fact that a failure or service discontinuity occurred;
- *Minimal:* The resumption of the regular operation;
- *Basic:* Type of failure or service discontinuity.

Likewise, ISO/IEC 15408–2 contains the following component definition for FAU_GEN.1:[20]

FAU_GEN.1.1 *The TSF shall be able to generate an audit record of the following auditable events:*
- *start-up and shutdown of the audit functions,*
- *all auditable events for the [selection: minimum, basic, detailed, not specified] level of audit; and*
- *[assignment: other specifically defined auditable events]*

The PP author decides to use the basic audit capability with no additional auditable events and specifies the requirement accordingly:

FAU_GEN.1.1 *The TSF shall be able to generate an audit record of the following auditable events:*
- *start-up and shutdown of the audit functions,*
- *all auditable events for the basic level of audit; and*
- *no other specifically defined auditable events.*

This requirement means that the developer would have to include the fact that a failure or service discontinuity occurred, how regular operations were resumed, and the type of failure or service discontinuity that occurred in the audit generation capability. Finally, two important points should be highlighted: (1) audit requirements are not specified with the FPT_RCV.2 component, but only with the FAU_GEN.1.1 element; and (2) FAU_GEN.1.1 is a good candidate for iteration because many functional components have potential auditable events identified in Part 2 of the CC standard. Audit requirements may also be specified for explicit requirements.

Operations are another area that require the CC standard to be accurately interpreted and applied. The CC methodology permits a degree of flexibility when expressing SFRs. This allows SFRs to be customized to meet the unique needs of a particular system or user community. The customization process, referred to as operations, is used to amplify requirements to the level of detail necessary. For example, operations may prescribe or forbid the use of particular security mechanisms.[19] Assignment and selection operations cannot be performed on all SFRs, only those so designated in ISO/IEC 15408-2. Care should be taken when performing operations so that no new dependencies are introduced. It is important to note that while hierarchies and dependencies relate to components, operations are performed on elements. Four types of operations may be performed on SFRs:

1. *Assignment* — Specify a parameter that is filled in when an element is used in a PP.[20]
2. *Selection* — Select one or more items from a list given in the CC Part 2 element definition.[20]
3. *Iteration* — Use an element more than once in a PP with varying parameters.[20]
4. *Refinement* — Add extra details, not found in the CC Part 2 definition, to an element when it is used in a PP or ST.[20]

The use of operations is illustrated in the four examples below.

The first example illustrates how the assignment operation works. ISO/IEC 15408-2 contains an SFR that states:[20]

FPT_RCV.2.2 *For* [**assignment:** list of failures/service discontinuities], *the TSF shall ensure the return of the TOE to a secure state using automated procedures.*

The square brackets indicate that an operation is not only permitted but also required. The **bold** letters clarify that the type of operation required is an assignment. The *italic* letters specify what parameters the PP author must supply. A PP author responds to this requirement by "filling in" the information requested between the square brackets, as appropriate for their specific system and security objectives. One possible response is below:

FPT_RCV.2.2 *For the partial or total loss of the network management function, security management function, and or the network transport capability the TSF shall ensure the return of the TOE to a secure state using automated procedures.*

The second example illustrates how the selection operation works. ISO/IEC 15408-2 contains an SFR that states:[20]

FTA_TAH.1.1 *Upon successful session establishment, the TSF shall display the* [**selection:** date, time, method, location] *of the last successful session establishment to the user.*

The square brackets indicate that an operation is not only permitted but also required. The **bold** letters clarify that the type of operation required is a selection. The *italic* letters specify the valid parameters from which the PP author must choose; no other parameters can be added. A PP author responds to this requirement by selecting the parameter or parameters that are appropriate for their specific system and security objectives. One possible response is below:

FTA_TAH.1.1	*Upon successful session establishment, the TSF shall display the date and time of the last successful session establishment to the user.*

Iteration consists of the repetitive use of the same element to address different aspects of a requirement.[101] Iteration can be performed on any functional element. A style convention has been developed to denote iteration: the use of a "+" sign and a number representing the iteration number (e.g., FCS_COP.1.1+2).[100,101] The third example illustrates how the iteration operation works in conjunction with assignment. ISO/IEC 15408-2 contains an SFR that states:[20]

FCS_COP.1.1	*The TSF shall perform* [**assignment:** list of cryptographic operations] *in accordance with a specified cryptographic algorithm* [**assignment:** cryptographic algorithm] *and cryptographic key sizes* [**assignment:** cryptographic key sizes] *that meet the following:* [**assignment:** list of standards].

A PP author could respond to this requirement by iterating the element to satisfy all cryptographic operations and completing the assignments. One possible response is:

FCS_COP.1.1+1	*The TSF shall perform digital signatures in accordance with a Level 2 cryptographic algorithm and Level 2 cryptographic key sizes that meet FIPS 140-2 requirements.*
FCS_COP.1.1+2	*The TSF shall perform message digests in accordance with a Level 2 cryptographic algorithm and Level 2 cryptographic key sizes that meet FIPS 140-2 requirements.*
FCS_COP.1.1+3	*The TSF shall perform encryption of user data during transmission in accordance with a Level 2 cryptographic algorithm and Level 2 cryptographic key sizes that meet FIPS 140-2 requirements.*

Refinement is an operation that allows customers to tailor a CC requirement to meet their specific needs. Refinement can be performed on any functional element. Refinement should only be used to provide an elaboration or specific interpretation of a CC requirement; it cannot be used to impose entirely new requirements (explicit requirements should be used for that purpose). The following example illustrates how the refinement operation works in conjunction with the selection example used earlier:

FTA_TAH.1.1	*Upon successful session establishment, the TSF shall display the date (dd-mm-yyyy) and time (in 24 hour military notation) of the last successful session establishment to the user.*

After the correct component has been selected, hierarchies evaluated, dependencies resolved, audit requirements specified, and element operations performed, the SFR is invoked in the PP. All constituent elements of a component are required to be used; there is no selection process for elements. The process shown in Exhibit 21 is repeated until all security objectives have been implemented through one or more SFRs.

3.6.2 Security Assurance Requirements (SARs)

Security assurance requirements (SARs) define the criteria for evaluating PPs, STs, and TOEs and the security assurance responsibilities and activities of developers and evaluators. SARs undergo a selection process similar to that for SFRs, as shown in Exhibit 25. The first step is to determine the level of protection needed. This contrasts with the selection of SFRs, which is based on the type of protection required to satisfy a security objective. The selection of SARs is driven by several factors:[21,22]

- Value of the assets to be protected versus the perceived risk of compromise
- Technical feasibility
- Development and evaluation costs/constraints
- Development and evaluation time requirements/constraints
- Current IT marketplace (COTS versus custom products)
- SFR to SAR and SAR to SAR dependencies

Information from this analysis is used to ascertain the level of security assurance needed (i.e., the degree of confidence that a TOE meets is security objectives).[19] Security assurance levels represent a continuum, from low to medium to high assurance. The goal is to ensure that a TOE is not over- or under-protected and to balance the level of assurance against technical feasibility, cost and schedule constraints, and need. It is useful to have a table at the beginning of this subsection that summarizes the applicable assurance components and indicates whether they are augmented and extended.

After the appropriate level of assurance has been identified, the corresponding evaluation assurance level is selected. Part 3 of ISO/IEC 15408 defines seven hierarchical EALs, with EAL 1 being the lowest level of assurance and EAL 7 the highest. An EAL is a grouping of assurance components — an assurance package. Each of the seven EALs adds new or higher assurance components that increase the scope, depth, and rigor of the evaluation as security objectives become more rigorous.

After an EAL has been selected, a determination is made whether or not this package is sufficient by itself for the specific situation to which it will be applied. If the standard EAL package is sufficient, it is used "as is" and the PP Identification field, 1.1.6 Common Criteria Conformance Claim and Version, is listed as "Part 3 conformant". If not, the PP author has two options for correcting deficiencies in the standard EAL package: augmentations and extensions. Use of one of the predefined EALs is not mandatory.[21] To illustrate:[96]

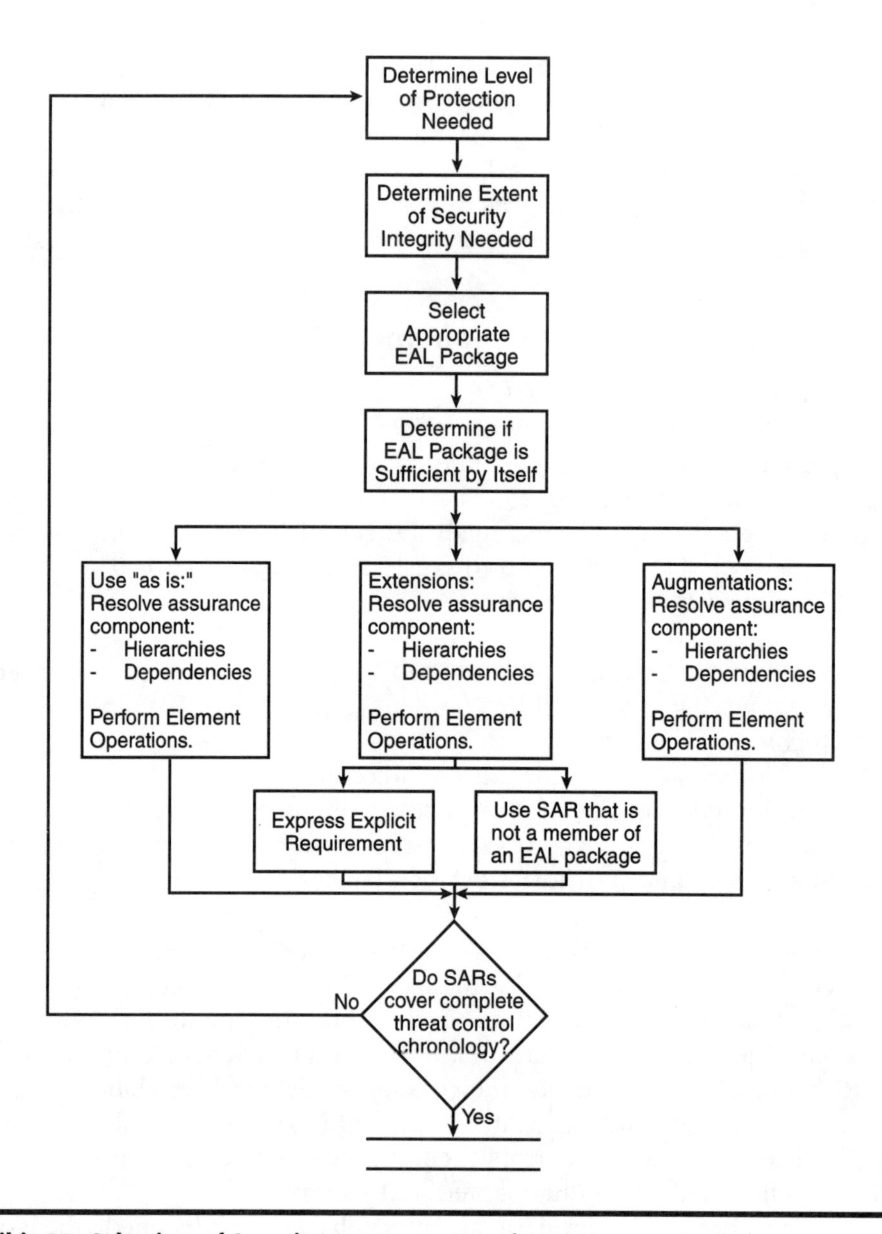

Exhibit 25. Selection of Security Assurance Requirements

> There is nothing sacred or magic about the EALs, and the PP author is free to specify alternative approaches, either by augmenting an existing EAL, or by developing an entirely new assurance package. This flexibility creates the opportunity to think carefully about building a cost-effective approach to evaluations and selecting components to address threats that exist for particular types of products [and systems].

Augmentations involve adding one or more standard assurance components from Part 3 of the CC to a predefined EAL. A higher component can be specified in lieu of one contained in the EAL package or a component can be specified from a family that is not contained in the EAL package.[101] The need for augmentation arises when a standard EAL package is almost right for a given situation but requires stronger measures in one or two assurance classes (configuration management, vulnerability assessment, and so forth). As an example, assume you have selected EAL 3; ACM_AUT.1 (partial CM automation) is normally an EAL 4 requirement. The TSF specified in your PP will be implemented in a multi-vendor environment; hence, it might be wise to augment EAL 3 with ACM_AUT.1. Augmentations are fairly common. In this case, the PP Identification field, 1.1.6 Common Criteria Conformance Claim and Version, is listed as "Part 3 augmented".

Extensions involve adding one or more assurance components to a predefined EAL that are not part of a standard EAL or are not part of ISO/IEC 15408-3:[96]

> The CC is designed to be extensible and it is possible to define functional and assurance requirements not contained in the CC. …Extensions may require prior approval of the National Evaluation Authority.

An EAL can be extended by one of two methods: specifying unused SARs and defining explicit SARs. As shown in Exhibit 26, 28 security assurance components are not included as part of a standard EAL. The CCIMB would not have taken the time to define these components if they were not intended to be used. These 28 components fall into four categories:

1. Those used to evaluate a PP
2. Those used to evaluate an ST
3. Those used to maintain an EAL between certification cycles
4. Those used to evaluate a TOE

An accredited CCTL uses the six assurance components in the APE class to evaluate a PP. These components are not used to augment an EAL, as EALs apply to TOEs. Likewise, the eight assurance components in the ASE class are used by a CCTL to evaluate an ST. These components are also not used to augment an EAL. The five assurance components in the AMA class are designed to ensure that an EAL is maintained after initial TOE certification and between certification cycles. These components should be included as part of the security assurance requirements in two scenarios:

1. The TOE will undergo repetitive certification cycles, such as triannual certification and accreditation (C&A).
2. The developer is a system integrator who will also be operating and maintaining the TOE.

Exhibit 26. Assurance Components That Are Not a Member of an EAL Assurance Package

Class	Family/Component	Purpose	When Used
ADO Delivery and Operation	ADO_IGS.2 Generation Log	Add requirement beyond ADO_IGS.1 for secure installation, generation, and start-up procedures to include a log of the options used to generate the TOE.	TOE evaluation
ADV Development	ADV_LLD.3 Formal Low-level Design	Add requirement beyond ADV_LLD.2 for the presentation of the low-level design to be formal.	TOE evaluation
	ADV_SPM.2* Semiformal TOE Security Policy Model	Add requirement beyond ADV_SPM.1 but less than ADV_SPM.3 for the presentation of the TOE security policy model to be semi-formal.	TOE evaluation
ALC Lifecycle Support	ALC_FLR.1 Basic Flaw Remediation	Require procedures for tracking, analyzing, and resolving security flaws to be documented.	TOE evaluation
	ACL_FLR.2 Flaw Reporting Procedures	Add requirement beyond ALC_FLR.1 for ensuring that reported flaws are resolved and resolved correctly so that no new flaws are introduced.	TOE evaluation
	ALC_FLR.3 Systematic Flaw Remediation	Add requirement beyond ALC_FLR.2 for designating a point of contact for reporting security flaws and ensuring their timely resolution.	TOE evaluation
AMA Assurance Maintenance	AMA_AMP.1 Assurance Maintenance Plan	Define policies and procedures for maintaining the security integrity of a TOE after it has been certified.	after initial TOE certification
	AMA_CAT.1 TOE Component Categorization Report	Rank order TOE components according to their relevance to security (this information is used by the AMA_AMP and AMA_SIA components).	after initial TOE certification
	AMA_EVD.1 Evidence of Maintenance Process	Define policies and procedures for generating, reporting, and storing evidence that assurance maintenance activities have been conducted and the results observed.	after initial TOE certification
	AMA_SIA.1 Sampling of Security Impact Analysis	Evaluate the security impact of proposed changes to the TOE after certification.	after initial TOE certification
	AMA_SIA.2 Examination of Security Impact Analysis	Evaluate the security impact of all proposed changes to the TOE after certification.	after initial TOE certification
APE Protection Profile Evaluation	APE_DES.1/Protection Profile, TOE Description, Evaluation Requirements	Determine if the TOE description is correct, complete, consistent, and coherent.	PP evaluation

Exhibit 26. Assurance Components That Are Not a Member of an Evaluation Assurance Level (EAL) Assurance Package (continued)

Class	Family/Component	Purpose	When Used
	APE_ENV.1 Protection Profile, Security Environment, Evaluation Requirements	Determine if the description of the TOE security environment is correct, complete, consistent, and coherent.	PP evaluation
	APE_INT.1 Protection Profile PP Introduction, Evaluation Requirements	Determine if the PP introduction is correct, complete, consistent, and coherent.	PP evaluation
	APE_OBJ.1 Protection Profile, Security Objectives, Evaluation Requirements	Determine if the security objectives are correct, complete, consistent, and coherent.	PP evaluation
	APE_REQ.1 Protection Profile), IT Security Requirements, Evaluation Requirements	Determine if the security requirements are correct, complete, consistent, and coherent.	PP evaluation
	APE_SRE.1 Protection Profile), Explicitly Stated IT Security Requirements, Evaluation Requirements	Determine if explicitly stated security requirements are correct, complete, consistent, and coherent.	PP evaluation
ASE Security Target Evaluation	ASE_DES.1 Security Target, TOE Description, Evaluation Requirements	Determine if the TOE description is correct, complete, consistent, and coherent.	ST evaluation
	ASE_ENV.1 Security Target, Security Environment, Evaluation Requirements	Determine if the description of the TOE security environment is correct, complete, consistent, and coherent.	ST evaluation
	ASE_INT.1 Security Target, ST Introduction, Evaluation Requirements	Determine if the ST introduction is correct, complete, consistent, and coherent.	ST evaluation
	ASE_OBJ.1 Security Target, Security Objectives, Evaluation Requirements	Determine if the security objectives are correct, complete, consistent, and coherent.	ST evaluation
	ASE_PPC.1 Security Target, PP Claims, Evaluation Requirements	Determine if the PP claims are correct.	ST evaluation
	ASE_REQ.1 Security Target, IT Security Requirements, Evaluation Requirements	Determine if the security requirements are correct, complete, consistent, and coherent.	ST evaluation
	ASE_SRE.1 Security Target, Explicitly Stated IT Security Requirements, Evaluation Requirements	Determine if explicitly stated security requirements are correct, complete, consistent, and coherent.	ST evaluation
	ASE_TSS.1 Security Target, TOE Summary Specification, Evaluation Requirements	Determine if the TOE Summary Specification is correct, complete, consistent, and coherent.	ST evaluation

Exhibit 26. Assurance Components That Are Not a Member of an Evaluation Assurance Level (EAL) Assurance Package (continued)

Class	Family/Component	Purpose	When Used
AVA Vulnerability Assessment	AVA_CCA.1 Covert Channel Analysis	Identify covert channels through an informal analysis.	TOE evaluation
	AVA_CCA.3 Exhaustive Covert Channel Analysis	Identify covert channels through an exhaustive analysis.	TOE evaluation
	AVA_VLA.2 Independent Vulnerability Analysis	Require that the vulnerability analysis be conducted by an independent organization.	TOE evaluation

* ADV_SPM.1 and ADV_SPM.3 are used, but not ADV_SPM.2.

The remaining nine assurance components are used to evaluate a TOE during either the initial evaluation or recertification. They can be used to augment an EAL package when appropriate. It is recommended that PP authors include one of the ALC_FLR components because flaw remediation is essential to maintaining a robust security posture.

The same process is used to specify explicit SARs as explicit SFRs. Explicit assurance components are modeled after standard assurance components and contain developer action elements, content and presentation evidence elements, and evaluator action elements. Explicit assurance components can be added to an existing assurance family, or a new assurance family or class can be developed. The following example illustrates the definition of a new assurance family. This family will be particularly useful during the ongoing operations and maintenance phase between certification cycles.

Explicit Requirement

ALC_SIR Security Incident Reporting

ALC_SIR.1 Basic security incident reporting

Dependencies:

No dependencies.

Developer action elements:

ALC_SIR.1.1D *Security incidents shall be reported to the customer.*

Content and presentation of evidence elements:

ALC_SIR.1.1C *Security incident reporting documentation shall describe the procedures for reporting security incidents to the customer.*

ALC_SIR.1.2C *Security incident reporting procedures shall include a description of what information will be reported.*

ALC_SIR.1.3C *Security incident reporting procedures shall include a description of how information will be reported.*

ALC_SIR.1.4C *Security incident reporting procedures shall include a description of when the information will be reported.*

Evaluator action elements:

ALC_SIR.1.1E	*The evaluator shall confirm that the information provided meets all requirements for content and presentation of evidence.*

ALC_SIR.2 Advanced security incident reporting

Dependencies:

No dependencies.

Developer action elements:

ALC_SIR.2.1D	***All physical, personnel, and information*** *security incidents shall be reported to the customer.*
ALC_SIR.2.2D	*All security incidents shall be reported to the customer, including known or suspected incidents, false positives, and false negatives.*

Content and presentation of evidence elements:

ALC_SIR.2.1C	*Security incident reporting documentation shall describe the procedures for reporting security incidents to the customer.*
ALC_SIR.2.2C	*Security incident reporting procedures shall include a description of what information will be reported.*
ALC_SIR.2.3C	*Security incident reporting procedures shall include a description of how information will be reported.*
ALC_SIR.2.4C	*Security incident reporting procedures shall include a description of when the information will be reported.*
ALC_SIR.2.5C	***Security incident reports shall be classified according to type, source, and severity.***

Evaluator action elements:

ALC_SIR.2.1E	*The evaluator shall confirm that the information provided meets all requirements for content and presentation of evidence.*

Part of determining whether or not an EAL package must be augmented or extended involves ensuring that SARs cover all phases of the threat-control chronology. SFRs and SARs correspond to the threat-control chronology differently. SFRs specify functions that the "as built" TOE will perform to anticipate, prevent, detect, characterize, and respond and recover from threats. In contrast, SARs specify developer and evaluator action elements, the performance of which will anticipate or prevent, detect or characterize, and respond and recover from vulnerabilities in any CC artifact (the PP, ST, or "as built" TOE) that could cause a threat to be carried out. The majority of SARs are preventive in nature. However, some SARs such as ADO_DEL, ADV_FSP, ALC_FLR, and ATE_IND go beyond prevention. Exhibit 27 maps SARs to security objectives.

Exhibit 27. Security Assurance Requirements (SARs) Mapped to Security Objectives

		Security Objective				
		Prevent	Detect		Correct	
SAR Class	SAR Family	Anticipate, Prevent	Detect	Characterize	Respond, Contain Consequences	Recover
ACM	AUT	X				
	CAP	X				
	SCP	X				
ADO	DEL	X	X			
	IGS	X				
ADV	FSP	X	X	X		
	HLD	X				
	IMP	X				
	INT	X				
	LLD	X	X	X		
	RCR	X				
	SPM	X				
AGD	ADM	X				
	USR	X				
ALC	DVS	X				
	FLR			X	X	X
	LCD	X				
	TAT	X				
ATE	COV	X				
	DPT	X				
	FUN	X	X	X		
	IND	X	X	X		
AVA	CCA	X	X	X		
	MSU	X			X	X
	SOF	X	X			
	VLA	X	X			
AMA	AMP	X				
	CAT	X				
	EVD	X	X			
	SIA	X	X	X		

The last step is to resolve dependencies; this process is quite similar to that used to resolve functional dependencies. Like SFRs, SARs may have internal or external, direct or indirect dependencies. While some SFRs have dependencies on SARs, no SARs have dependencies on SFRs. Also, SARs do not have any dependencies where the PP author has the option to choose from a list of components to satisfy the dependency. Assurance components have one feature that functional components do not — that of reciprocal dependencies. If SAR *A* is dependent on SAR *B*, and SAR *B* is also dependent on SAR *A*, that is referred to as a reciprocal dependency. Exhibit 28 lists the assurance dependencies.

The only operations permitted on assurance elements are refinement and iteration. These operations are performed the same as for SFRs. Iteration is used in situations where a component is applicable to: (1) one or more FPs in a component TOE with different levels of refinement, or (2) one or more TOEs in a composite TOE, again with different levels of refinement. Assignment and selection operations are not allowed.

Finally, if the SFRs include any requirements for identification, authentication, or encryption functions based on statistical or quantitative mechanisms or an EAL greater than EAL 2 has been selected, the AVA_SOF.1 assurance component must be specified.

3.6.3 Security Requirements for the IT Environment

Security requirements are specified for the IT environment in which the TOE will operate. This subsection captures IT functions on which the TOE is dependent; the absence of these functions may cause the TOE to operate insecurely. Requirements for the IT environment can be stated to reinforce or strengthen environmental assumptions made in Section 3 of the PP or in response to security objectives for the environment. Like SFRs, Requirements for the IT environment can be stated using standard requirements from Part 2 of the CC or explicit requirements. Again, hierarchies must be evaluated, dependencies resolved, and operations performed. If the TOE is not dependent on the IT environment, that should be stated as well. The examples below illustrate security requirements for the IT environment. They are derived from the sample security objectives for the environment presented in Exhibit 20. The two explicit requirements reflect security objectives O15 and O16. The conformant requirements reflect security objective O18 and O19.

Example 5.3: Security Requirements for the IT Environment

Explicit requirement

FCS_CIF.1 Cryptographic infrastructure

FCS_CIF.1.1	*The TOE environment shall support the enterprise-wide cryptographic infrastructure.*

Exhibit 28. Assurance Dependencies[21]

Class/ Component	None	Internal Dependencies		Dependency with Options	External Dependencies	
		Direct	Indirect		Direct	Indirect
APE_DES.1	—	APE_ENV.1 **APE_INT.1** APE_OBJ.1 APE_REQ.1	—	—	—	—
APE_ENV.1	X	—	—	—	—	—
APE_INT.1	—	**APE_DES.1** APE_ENV.1 APE_OBJ.1 APE_REQ.1	—	—	—	—
APE_OBJ.1	—	APE_ENV.1	—	—	—	—
APE_REQ.1	—	APE_OBJ.1	—	—	—	—
APE_SRE.1	—	APE_REQ.1	—	—	—	—
ASE_DES.1	—	ASE_ENV.1 **ASE_INT.1** ASE_OBJ.1 ASE_PPC.1 ASE_REQ.1 ASE_TSS.1	—	—	—	—
ASE_ENV.1	X	—	—	—	—	—
ASE_INT.1	—	**ASE_DES.1** ASE_ENV.1 ASE_OBJ.1 ASE_PPC.1 ASE_REQ.1 ASE_TSS.1	—	—	—	—
ASE_OBJ.1	—	ASE_ENV.1	—	—	—	—
ASE_PPC.1	—	ASE_OBJ.1 ASE_REQ.1	—	—	—	—
ASE_REQ.1	—	ASE_OBJ.1	—	—	—	—
ASE_SRE.1	—	ASE_REQ.1	—	—	—	—
ASE_TSS.1	—	ASE_REQ.1	—	—	—	—
ACM_AUT.1 ACM_AUT.2	—	ACM_CAP.3	ACM_SCP.1	—	—	ALC_DVS.1
ACM_CAP.1 ACM_CAP.2	X	—	—	—	—	—
ACM_CAP.3 ACM_CAP.4	—	—*	—	—	ALC_DVS.1	—
ACM_CAP.5	—	—*	—	—	ALC_DVS.2	—
ACM_SCP.1 ACM_SCP.2 ACM_SCP.3	—	ACM_CAP.3	—	—	—	ALC_DVS.1
ADO_DEL.1	X	—	—	—	—	—
ADO_DEL.2 ADO_DEL.3	—	ACM_CAP.3	ACM_SCP.1	—	—	ALC_DVS.1
ADO_IGS.1 ADO_IGS.2	—	—	—	—	AGD_ADM.1	ADV_FSP.1 ADV_RCR.1

Exhibit 28. Assurance Dependencies[21] (continued)

Class/ Component	None	Internal Dependencies		Dependency with Options	External Dependencies	
		Direct	Indirect		Direct	Indirect
ADV_FSP.1 ADV_FSP.2 ADV_FSP.3 ADV_FSP.4	X	ADV_RCR.1	—	—	—	—
ADV_HLD.1 ADV_HLD.2	—	ADV_FSP.1 ADV_RCR.1	—	—	—	—
ADV_HDL.3 ADV_HDL.4	—	ADV_FSP.3 ADV_RCR.2	—	—	—	—
ADV_HDL.5	—	ADV_FSP.4 ADV_RCR.3	—	—	—	—
ADV_IMP.1 ADV_IMP.2	—	ADV_LLD.1 ADV_RCR.1	ADV_FSP.1 ADV_HLD.2	—	**ALC_TAT.1**	—
ADV_IMP.3	—	ADV_INT.1 ADV_LLD.1 ADV_RCR.1	ADV_FSP.1 ADV_HLD.2	—	ALC_TAT.1	—
ADV_INT.1 ADV_INT.2	—	ADV_IMP.1 ADV_LLD.1	ADV_FSP.1 ADV_HLD.2 ADV_RCR.1	—	ALC_TAT.1	—
ADV_INT.3	—	ADV_IMP.2 ADV_LLD.1	ADV_FSP.1 ADV_HLD.2 ADV_RCR.1	—	ALC_TAT.1	—
ADV_LLD.1	—	ADV_HLD.2 ADV_RCR.1	ADV_FSP.1	—	—	—
ADV_LLD.2	—	ADV_HLD.3 ADV_RCR.2	ADV_FSP.3	—	—	—
ADV_LLD.3	—	ADV_HLD.5 ADV_RCR.3	ADV_FSP.4	—	—	—
ADV_RCR.1 ADV_RCR.2 ADV_RCR.3	X	—	—	—	—	—
ADV_SPM.1 ADV_SPM.2 ADV_SPM.3	—	ADV_FSP.1	ADV_RCR.1	—	—	—
AGD_ADM.1	—	—	—	—	ADV_FSP.1	ADV_RCR.1
AGD_USR.1	—	—	—	—	ADV_FSP.1	ADV_RCR.1
ALC_DVS.1 ALC_DVS.2	X	—	—	—	—	—
ALC_FLR.1 ALC_FLR.2 ALC_FLR.3	X	—	—	—	—	—
ALC_LCD.1 ALC_LCD.2 ALC_LCD.3	X	—	—	—	—	—
ALC_TAT.1 ALC_TAT.2 ALC_TAT.3	—				**ADV_IMP.1**	ADV_FSP.1 ADV_HLD.2 ADV_LLD.1 ADV_RCR.1

Exhibit 28. Assurance Dependencies[21] (continued)

Class/ Component	None	Internal Dependencies		Dependency with Options	External Dependencies	
		Direct	Indirect		Direct	Indirect
ATE_COV.1 ATE_COV.2 ATE_COV.3	—	ATE_FUN.1	—	—	ADV_FSP.1	ADV_RCR.1
ATE_DPT.1	—	ATE_FUN.1	—	—	ADV_HLD.1	ADV_FSP.1 ADV_RCR.1
ATE_DPT.2	—	ATE_FUN.1	—	—	ADV_HLD.2 ADV_LLD.1	ADV_FSP.1 ADV_RCR.1
ATE_DPT.3	—	ATE_FUN.1	—	—	ADV_HLD.2 ADV_IMP.2 ADV_LLD.1	ADV_FSP.1 ADV_RCR.1 ALC_TAT.1
ATE_FUN.1 ATE_FUN.2	X	—	—	—	—	—
ATE_IND.1	—	—	—	—	ADV_FSP.1 AGD_ADM.1 AGD_USR.1	ADV_RCR.1
ATE_IND.2 ATE_IND.3	—	ATE_FUN.1	—	—	ADV_FSP.1 AGD_ADM.1 AGD.USR.1	ADV_RCR.1
AVA_CCA.1 AVA_CCA.2 AVA_CCA.3	—	—	—	—	ADV_FSP.2 ADV_IMP.2 AGD_ADM.1 AGD_USR.1	ADV_HLD.2 ADV_LLD.1 ADV_RCR.1 ALC_TAT.1
AVA_MSU.1 AVA_MSU.2 AVA_MSU.3	—	—	—	—	ADO_IGS.1 ADV_FSP.1 AGD_ADM.1 AGD_USR.1	ADV_RCR.1
AVA_SOF.1	—	—	—	—	ADV_FSP.1 ADV_HLD.1	ADV_RCR.1
AVA_VLA.1	—	—	—	—	ADV_FSP.1 ADV_HLD.1 AGD_ADM.1 AGD_USR.1	ADV_RCR.1
AVA_VLA.2 AVA_VLA.3 AVA_VLA.4	—	—	—	—	ADV_FSP.1 ADV_HLD.2 ADV_IMP.1 ADV_LLD.1 AGD_ADM.1 AGD_USR.1	ADV_RCR.1 ALC_TAT.1
AMA_AMP.1	—	AMA_CAT.1	—	—	ACM_CAP.2 ALC_FLR.1	—
AMA_CAT.1	—	—	—	—	ACM_CAP.2	—
AMA_EVD.1	—	AMA_AMP.1 AMA_SIA.1	AMA_CAT.1	—	—	—
AMA_SIA.1 AMA_SIA.2	—	AMA_CAT.1				

Note: **Bold** indicates a reciprocal dependency.

*Per Final Interpretation 065.

Explicit requirement

FPT_ENV.1, Environmental Protection

FPT_ENV.1.1	*The TOE shall monitor for and provide protection against natural and manmade environmental threats (fire, flood, humidity, dust, vibration, earthquakes, temperature fluctuation, power fluctuations, and so forth).*

FPT_PHP.3 Resistance to physical attack

FPT_PHP.3.1	*The TSF shall resist malicious physical attacks, tampering, unauthorized modification, destruction, and theft to TSF by responding automatically such that the TSP is not violated.*

FPT_SSP.2 Mutual trusted acknowledgment

FPT_SSP.2.1	*The TSF shall acknowledge, when requested by another part of the TSF, the receipt of an unmodified TSF data transmission.*
FPT_SSP.2.2	*The TSF shall ensure that the relevant parts of the TSF know the correct status of transmitted data among its different parts, using acknowledgments.*

FPT_STM.1 Reliable time stamps

FPT_STM.1.1	*The TSF shall be able to provide reliable time stamps for its own use.*

3.6.4 Security Requirements for the Non-IT Environment

Security requirements may also be specified for the non-IT environment in which the TOE will operate. Requirements for the non-IT environment can be stated to reinforce or strengthen assumptions made in Section 3 of the PP about the intended use and operation of the TOE, organizational security policies, and in response to security objectives for the environment. Like SFRs, requirements for the non-IT environment can be stated using standard requirements from Part 2 of the CC or explicit requirements. Again, hierarchies must be evaluated, dependencies resolved, and operations performed. If the TOE has no dependencies on the non-IT environment, that should be stated as well. The examples that follow illustrate the security requirements for the non-IT environment. They are derived from the sample security objectives (O13and O14) for the environment presented in Exhibit 20.

Example 5.4: Security Requirements for the Non-IT Environment

Explicit Requirement

FAU_SAP.1, Security Audit Processing

FAU_SAP.1.1	*System activity audit records shall be: (a) reviewed daily by authorized users and administrators; (b) stored online for 7 days; and (c) stored offline for 90 days.*

Explicit requirement

FMT_ACR.1, Access Control Rights and Privileges

FMT_ACR.1.1	*The system security administrator shall implement access control rights and privileges within 15 minutes of being directed to do so.*

3.7 Section 6: PP Application Notes

PP Application Notes are optional. They provide an opportunity for a customer to convey additional background information to potential developers to help them understand the security problem being solved and interpret SFRs correctly. PP Application Notes may be interspersed throughout Section 5.1 as needed or collected in Section 6. If no PP Application Notes are to be added, then a statement should be made to that effect in Section 6. Exhibit 29 illustrates the three different methods for expressing PP Application Notes. Information contained in PP Application Notes does not carry the weight of a requirement, is not binding on a developer, and is not evaluated by an evaluator. As a result, the use of PP Application Notes should be kept to a minimum.

3.8 Section 7: Rationale

Section 7 of a PP, the Rationale, sets a PP apart from most other types of requirements specifications. The Rationale proves that the requirements specified are complete, coherent, consistent, and correct. In contrast, other methodologies only state requirements and optionally convey background information. To generate this logical proof, PP assumptions, threats, policies, objectives, and requirements are subjected to a rigorous analytical process. This process is similar to a formal mathematical proof and formal methods, in general. Cost, schedule, and technical benefits from generating a PP Rationale are significant. Several studies conducted during the last decade, inside and outside the United States, concluded that ~85 percent of the failures or

Exhibit 29. PP Application Notes Example

Example 1: Separate Section — No application notes

Section 6 PP Application Notes

This PP does not contain any optional PP Application Notes.

Example 2: Separation Section — With application notes

Section 6 PP Application Notes

5.1.3.1 1. User Identification Before Any Action (FIA_UAU.2)

For this requirement the term "user" refers to source of transmitted information, whether humans or processes.

5.1.3.4 4. Cryptographic Operation (FCS_COP.1.1)

Future migration to the advanced encryption algorithm (AES) is anticipated and will be approved when products are available.

Example 3: Interspersed Application Notes

5.1.3.1 1. User Identification Before Any Action (FIA_UAU.2)

The TSF shall require each user to be successfully authenticated before allowing any other TSF-mediated actions on behalf of that user.

Application Note: For this requirement the term "user" refers to source of transmitted information, whether humans or processes.

5.1.3.4 4. Cryptographic Operation (FCS_COP.1.1)

The TSF shall perform data encryption services in accordance with the 3DES cryptographic algorithm and cryptographic key sizes of 112 bits that meet the FIPS PUB 140–2, level 2 standard.

Application Note: Future migration to the Advanced Encryption Algorithm (AES) is anticipated and will be approved when products are available.

defects in IT systems were due to requirement errors. Taking the time to prove that requirements are correct prior to design and implementation saves a lot of potentially wasted time and resources; it is much easier, less expensive, and faster to fix an erroneous requirement found during the requirements analysis phase than one found during verification of an already built system.[99]

The purpose of the Rationale is to "demonstrate that a conformant TOE will provide an effective set of IT security countermeasures within the TOE security environment."[22] A PP Rationale consists of two subsections: Security Objectives Rationale and Security Requirements Rationale. A Rationale may be produced as a stand-alone section because it is primarily used by customers and evaluators, not developers.

3.8.1 Security Objectives Rationale

The Security Objectives Rationale presents evidence to demonstrate that stated security objectives are traceable and suitable to cover all aspects of the TOE security environment: assumptions, threats, and security policies.[22] The Security Objectives Rationale must prove that the security objectives are:

- *Necessary* — Each assumption, threat, and security policy must be encompassed by one or more security objectives. Any assumption, threat, or security policy that is not accounted for is flagged as an error requiring resolution.
- *Appropriate* — Each security objective must correlate to at least one assumption, threat, or security policy. Any extraneous security objectives are flagged as errors requiring resolution.
- *Sufficient* — The totality of security objectives must correspond to the threat control chronology: anticipate or prevent, detect or characterize, respond or recover. A mapping is performed to ensure that the security objectives adequately cover the entire threat control chronology. Any shortcomings are noted as errors requiring resolution.

Common practice is to convey this information in a table at the beginning of this subsection that correlates assumptions, threats, and policies with security objectives. Following the table a short discussion of each intersection is provided (see Exhibit 30). In this example, an assumption, threat, and policy correlate with three security objectives. In other cases, an assumption, threat, or policy alone may correlate to a security objective.

3.8.2 Security Requirements Rationale

The Security Requirements Rationale presents evidence to demonstrate that the combined security functional and assurance requirements for the TOE, the IT environment, and non-IT environment are traceable and suitable to satisfy all security objectives.[22] Six items must be justified or demonstrated:[22]

1. Use of explicit SFRs
2. Choice of an EAL
3. Need for EAL augmentations and extensions
4. Unresolved dependencies
5. Selection of a strength of functional level
6. Strength of function claim

The justification for each of these six items is provided in a few paragraphs of text. For example, the choice of an EAL is explained by the primary influential factors such as:

- Applicable laws, regulations, and policies
- Sophistication, motivation, and resources of potential attackers
- Value of assets to be protected
- Criticality of the mission performed by the TOE
- Complexity of system integration issues surrounding the TOE

Exhibit 30. Sample Security Objectives Rationale

Table x Summary of Security Objectives Rationale

Assumption	Threat	Security Policy	Security Objective
A1	T3	P2	O2, O3, O4

- A1 Users and administrators are restricted from importing untrusted code.
- T3 An authorized user or administrator of the TOE may unwittingly introduce malicious code into the system, resulting in a compromise of the integrity and/or availability of user and/or system resources.
- P2 The TSF limits the access control rights and privileges of users and administrators. S

Authorized users and administrators are trained about the secure installation and operation of the target of evaluation (TOE) (O2); TOE detects any unauthorized changes to configuration and operation (O3); TOE protects itself from insider and outsider attacks, such as those from malicious code (O4).

The security requirements rationale must also prove that the combination of SFRs and SARs are:

- *Necessary* — Each security objective must be encompassed by at least one SFR or SAR. Any security objective that is not accounted for is flagged as an error requiring resolution.
- *Appropriate* — Each SFR and SAR must correlate to at least one security objective. Any extraneous SFRs or SARs are noted as errors requiring resolution.
- *Sufficient* — SFRs and SARs must be complete, coherent, consistent, and correct. Hierarchies must have been evaluated correctly, all dependencies must have been resolved or the non-resolution adequately justified, audit requirements must be complete, and operations must have been performed correctly. SFRs must prevent tampering, deactivation, misconfiguration, and bypassing of TOE security functions.Common practice is to convey this information in a table at the beginning of this subsection that correlates security objectives to security requirements. Following the table a short discussion of each intersection is provided (see Exhibit 31). In this example one security objective correlates with two security requirements.

3.9 Summary

A PP is a formal document which expresses an *implementation-independent* set of security requirements, both functional and assurance, for an IT product or system that meets specific consumer needs.[19,23,110] The process of developing a PP guides a consumer to elucidate, define, and validate their security requirements, the end result of which is used to (1) communicate these requirements to potential developers, and (2) provide a foundation from which a Security Target can be developed and a formal evaluation conducted. Several stakeholders interact with a PP. PPs are written by customers (or end users), read by potential developers (vendors) and system integrators, and reviewed and assessed by evaluators. Once written, a PP is not cast in concrete; rather,

Exhibit 31. Sample Security Requirements Rationale

Table x: Summary of Security Requirements Rationale

Security Objective	Security Requirement
O1	FTA_TAB.1 FMT_MOF.1
• O1 The TOE will provide a banner to notify all users that they are entering a restricted government computer system. FTA_TAB.1 provides the capability to display warning banners to all users and system administrators logging onto the TOE; FMT_MOF.1 provides the capability for administrators to change or replace the text of the banner as necessary.	

it is a living document. The CC/CEM methodology and artifacts map to generic system lifecycle and procurement phases. A PP corresponds to the requirements analysis phase — customers state their IT security requirements in PPs and the quality of these security requirements is verified through the APE class security assurance activities. A PP is part of pre-award procurement activities; PPs are included in the requests for proposal made available to potential offerors.

A PP is a formal document with specific content, format, and syntax requirements. This formality is imposed to ensure that PPs are accurately and uniformly interpreted by all stakeholders. A PP is not written *per se*; rather, it captures the culmination of a series of analyses conducted by customers to elucidate, define, and validate their security requirements. A PP is a cohesive whole; as such, extensive interaction exists among the six required sections. Section 1 introduces a PP by identifying its nature, scope, and status. Section 2 describes the assets that require protection and the sensitivity of each and defines the TOE boundaries.

Section 3 states the assumptions, analyzes the threats, and cites organizational security policies that are applicable to the TSF. Assumptions about the intended use, operational environment, connectivity, and roles and responsibilities are articulated. Any environmental constraints or operational limitations are clarified. Potential threats are ascertained and itemized by the TOE. Then, the likelihood of each threat occurring is estimated and the severity of the consequences should the threat be carried out is determined. Because all threats are not equivalent, a risk mitigation priority is established for each potential threat predicated on its severity and likelihood. Organizational security policies include rules, procedures, and practices that an organization imposes on an IT system to protect its assets.[19] Local, national, or international laws and regulations may impose additional policies (for example, privacy requirements).

Section 4 delineates security objectives for the TOE and the IT environment. These objectives are derived from an analysis of the assumptions, threats, and security policies articulated in Section 3. Security objectives are written for the TOE and the operational environment (IT and non-IT). Countermeasures deployed by the TOE satisfy TOE security objectives.[22] Technical measures implemented by the IT environment meet security objectives for the IT environment,[22] while procedural measures achieve security objectives for the non-IT environment.[22]

Section 5 implements security objectives through a combination of SFRs and SARs. These SFRs and SARs are derived from an analysis of the sensitivity of the assets to be protected as stated in Section 2 and the perceived risk of compromise presented in Section 3. A systematic decision-making process is followed to select both SFRs and SARs, as illustrated in Exhibits 21 and 25. These processes include, among other activities, the evaluation of hierarchies, the resolution of dependencies, the statement of explicit requirements, the specification of audit requirements, and the performance of element operations.

Section 6, which is optional, provides an opportunity for a customer to relay additional background information to developers and evaluators.

The last section, Section 7, proves that requirements specified in Section 5 implement all security objectives stated in Section 4 for the security environment defined in Section 3. Section 7 of a PP, the Rationale, sets a PP apart from most other types of requirements specifications. The Rationale proves that the requirements specified are complete, coherent, consistent, and correct. In contrast, other methodologies only state requirements and optionally convey background information. To generate this logical proof, PP assumptions, threats, policies, objectives, and requirements are subjected to a rigorous analytical process. This process is similar to a formal mathematical proof and formal methods, in general. Significant cost, schedule, and technical benefits can be gained by generating a PP Rationale.

3.10 Discussion Problems

1. Which operations are legal for SFRs, but not SARs, and why? Which operations are legal for SARs, but not SFRs, and why?
2. Explain the difference between augmented and extended. Can a PP be both augmented and extended? Where is this explained in the PP?
3. What is the relationship between: (a) assumptions and threats; (b) SFRs, SARs, and objectives; and (c) policies and an EAL?
4. Why should dependencies be resolved? What is a possible reason for not doing so? What is the difference between a direct and indirect dependency?
5. Explain the difference between monolithic, component, and composite TOEs? Which type of TOE contains packages?
6. Why is the information captured by a CCRA participant PP registry different than that captured by the ISO/IEC JTC 1 RA?
7. What is the asset sensitivity level used for?
8. Describe the impact of TOE boundary definitions.
9. How does a developer use the assumptions stated in Section 2.1 of a PP?
10. How are threats stated in Section 2.2 of a PP characterized, and why?
11. Why are organizational security policies contained in a PP?
12. How are security objectives classified, and why?

13. How are audit requirements specified?
14. What is the difference between a direct and an indirect dependency? Can an SFR satisfy both?
15. Are standard CC SFRs considered explicit requirements? Explain why.
16. Create an EAL package that uses SARs that are not part of a standard EAL. Define the purpose of the package and how it relates to the EAL scale.
17. Explain the difference and similarities between functional and assurance dependencies.
18. How are explicit requirements developed?
19. What makes a PP unique compared to other requirements specifications?
20. Explain the difference and interaction between: (a) SFRs and security requirements for the IT environment, (b) SFRs and security requirements for the non-IT environment, and (c) security requirements for the IT and non-IT environments.
21. How and why is a minimum SOF stated? How is the selection of a SOF justified?
22. What determines whether or not the security objectives are sufficient? What determines whether or not the security requirements are sufficient?

Chapter 4

Designing a Security Architecture: The Security Target

This chapter explains how to design a security architecture, in response to a Protection Profile (PP), through the instrument of a Security Target (ST) using the Common Criteria (CC) standardized methodology, syntax, and notation. The required content and format of an ST are discussed section by section. The perspective from which to read and interpret STs is defined. In addition, the purpose, scope, and development of an ST are mapped to both a generic system lifecycle and a generic procurement sequence.

4.0 Purpose

A Security Target, a combination of security objectives, functional and assurance requirements, summary specifications, PP claims, and rationales, is an *implementation-dependent* response to a PP that is used as the basis for the development and evaluation of a target of evaluation (TOE).[19,24,112] In other words, the PP specifies security functional and assurance requirements, while an ST provides a detailed design that incorporates specific functional security mechanisms and security assurance measures to fulfill these requirements. It is possible for several different, yet equally valid, STs to be written in response to one PP because of the implementation dependence. The evaluation of an ST concentrates on verifying that it is an adequate, complete, correct, and consistent interpretation of the requirements stated in the applicable PP.

Several stakeholders interact with an ST. STs are written by developers in response to a PP, read by potential customers, and reviewed and assessed by evaluators. Once approved, STs are used by developers as the foundation for constructing a TOE. Certified STs are generally posted on the Web in registries maintained by the National Evaluation Authorities concurrent with the Evaluated Products List; STs are not posted until the target of evaluation (TOE) itself is certified. (Note that STs posted on Neb sites have often been sanitized to remove corporate proprietary or sensitive security information; hence, they may not be "complete.")

An ST should be self contained and oriented toward the main stakeholders — the customer and evaluator; readers should not have to refer back to a multitude of other documents. An ST presents a concise statement of how security functional and assurance requirements will be implemented. An ST is not cast in concrete, rather, it is a living document. Updates to an ST may be triggered by:[22]

- Identification of and response to new threats
- Changes in organizational security policies
- Changes in system mission or intended use
- New cost or schedule constraints
- Higher than expected development costs
- Changes in the allocation of requirements between (a) a TOE and its environment, or (b) component TOEs
- New technology
- Deficiencies uncovered during evaluation or operation of the TOE
- (Re)certification activities (Common Criteria and certification and accreditation [C&A])

A variety of system lifecycle models have been developed over the years, such as structured analysis and design, the classic waterfall model, step-wise refinement, spiral development, rapid application development/joint application development (RAD/JAD), object-oriented analysis and design, and formal methods. While the sequence, duration, and feedback among the phases of each of the models differ, they all contain certain generic lifecycle phases: concept, requirements analysis and specification, design, development, verification, validation, operations and maintenance, and decommissioning. The Common Criteria/Common Evaluation Methodology (CC/CEM) and artifacts are not tied to any specific lifecycle model; rather, they reflect a continuous process of refinement with built-in checks. As the standard states, the CC/CEM methodology is:[19]

> ...based on the refinement of the PP security requirements into a TOE Summary Specification expressed in the ST. Each lower level of refinement represents a further decomposition with additional design detail. The least abstract representation is the TOE implementation itself. ...The CC requirement is that there should be sufficient design representations presented at sufficient level of granularity to demonstrate: (1) that each refinement level is a complete instantiation of the higher levels (i.e., all TOE security functions, properties, and behavior defined

at the higher level of abstraction must be demonstrably present in the lower level), and (2) that each refinement level is an accurate instantiation of the higher levels (i.e. there should be no additional TOE security functions, properties or behavior defined at the lower level of abstraction that are not specified at the higher level).

While the CC/CEM does not dictate a particular lifecycle model, depending on the evaluation assurance level (EAL), the developer may have to justify the model followed. The ALC_LCD security assurance activities evaluate the appropriateness, standardization, and measurability of the lifecycle model used by the developer. ALC_LCD.1 (Developer-Defined Lifecycle Model) requires the developer to establish and use a lifecycle model for the development and maintenance of the TOE, including lifecycle documentation. ALC_LCD.2 (Standardized Lifecycle Model) adds requirements to explain why the model was chosen, how it is used to develop and maintain the TOE, and how documentation demonstrates compliance with the model. ALC_LCD.3 (Measurable Lifecycle Model) adds requirements to explain the metrics used to measure compliance with the lifecycle model during the development and maintenance of the TOE. EAL 4 includes ALC_LCD.1, EALs 5 and 6 include ALC_LCD.2, and EAL 7 includes ALC_LCD.3.

The CC/CEM and artifacts map to generic lifecycle phases. An ST corresponds to the design phase — a design is generated by the developer in response to security requirements stated by a customer in a PP and the quality of this design is verified through the ASE class security assurance activities.

Likewise, large system procurements go through a series of generic phases. Pre-award activities include concept definition, feasibility studies, independent cost estimates, the issuance of a Request for Proposals (RFPs), and proposal evaluation. Post-award activities include contract award; monitoring system development; accepting delivery orders; issuing engineering change orders (ECOs) to correct deficiencies in requirements, design, or development; and, finally, system deployment. After system roll-out and acceptance are complete, organizations generally transition to an operations and maintenance contract that lasts through decommissioning. An ST is part of pre-award procurement activities; STs are included in the proposals submitted by potential offerors and evaluated by the source selection team. This practice is becoming prevalent among government agencies in the United States, in part because NSTISSP #11 (National Information Assurance Acquisition Policy)**75** mandated the use of CC-evaluated information technology (IT) security products in critical infrastructure systems starting in July 2002. Exhibit 1 aligns CC/CEM artifacts and activities with generic system lifecycle phases and generic procurement phases.

4.1 Structure

An ST is a formal document with specific content, format, and syntax requirements. This formality is imposed to ensure that STs are accurately and uniformly interpreted by all the different stakeholders. An ST is not written

Exhibit 1. Mapping of CC/CEM Artifacts to Generic System Lifecycle and Procurement Phases

CC/CEM Artifacts and Activities	Generic System Lifecycle Phases	Generic Procurement Phases
none	Concept	• Concept definition • Feasibility studies, needs analysis • Independent cost estimate
• Protection Profile (PP) • Security assurance activity: APE	Requirements analysis and specification	Request for proposal (tender) issued by customer
• Security Target (ST) • Security assurance activity: ASE	Design	• Technical and cost proposals submitted by vendors • Technical and cost proposals evaluated by customer
• Target(s) of Evaluation (TOE) developed by winning vendor • Security assurance activities: ACM, ADV	Development	Contract award
Security assurance activities: ATE, AVA	Verification	• Acceptance of delivery orders • ECPs issued to correct deficiencies in requirements, design, or development
Security assurance activities: ADO, AGD	Validation, installation and checkout	Deployment
Security assurance activities: ALC, AVA, AMA	Operations and maintenance	Transition to maintenance contract
none	Decommissioning	Contract expiration

per se; rather, it captures the culmination of a series of analyses conducted by the developer to create a solution to a customer's security requirements. As shown in Exhibit 2, an ST consists of eight sections; all sections are required. The content and development of each of the eight sections are discussed in detail below in Sections 4.2 through 4.9. All information should be placed in the appropriate section in an ST where it will be found and correctly interpreted.

An ST is a cohesive whole; as such, interaction among the eight required sections is extensive. This underscores the importance of putting information in the correct section. Section 1 introduces an ST by identifying its nature, scope, and status. Section 2 describes the system type, architecture, and the logical and physical security boundaries. Section 3 states the assumptions, analyzes the threats, and cites organizational security policies applicable to the TSF. Section 4 delineates security objectives for the TOE and the IT environment. These objectives are derived from an analysis of the assumptions, threats, and security policies articulated in Section 3. Section 5 implements security objectives expressed in Section 4 through a combination of security functional requirements (SFRs) and security assurance requirements (SARs). These SFRs and SARs are derived from an analysis of the security architecture

Exhibit 2. Content of a Security Target (ST)

1. Security Target Identification
1.1 Security Target Identification
1.1.1 ST Name:
1.1.2 ST Identifier:
1.1.3 Keywords:
1.1.4 EALs:
1.1.5 ST Evaluation Status:
1.2 Security Target Overview
1.2.1 ST Overview
1.2.2 Strength of Function
1.2.3 Related PPs, STs, and Referenced Documents
1.2.3 ST Organization
1.2.4 Acronyms
1.3 ISO/IEC 15408 Conformance
2. Description
2.1 System Type
2.2 Architecture
2.3 Security Boundaries
2.3.1 Physical Boundaries
2.3.2 Logical Boundaries
3. Security Environment
3.1 Assumptions
3.1.1 Intended Use
3.1.2 Operational Environment
3.1.3 Connectivity
3.2 Threats
3.2.1. Threats Addressed by the TOE
3.2.2. Threats Addressed by the Operational Environment
3.3. Organizational Security Policies
4. Security Objectives
4.1. Security Objectives for the TOE
4.2. Security Objectives for the Operational Environment
4.2.1. IT Environment
4.2.2. Non-IT Environment
5. Security Requirements
5.1. Security Functional Requirements (SFRs)
5.2. Security Assurance Requirements (SARs)
5.3. Security Requirements for the IT Environment
5.4. Security Requirements for the Non-IT Environment
6. TOE Summary Specification
6.1. TOE Security Functions
6.1.*x*. [discussion of each functional package/class]
6.2. Assurance Measures
6.2.*x*. [discussion of each assurance package/class]
7. PP Claims
7.1. PP Reference
7.2. PP Tailoring
7.3. PP Additions
8. Rationale
8.1. Security Objectives Rationale
8.2. Security Requirements Rationale
8.3. TOE Summary Specification Rationale
8.4. PP Claims Rationale

and boundaries discussed in Section 2 and the perceived risk of compromise presented in Section 3. Section 6 describes the specific functional security mechanisms and security assurance measures employed to fulfill the requirements stated in Section 5. Section 7 clarifies which PP the ST was developed in response to and any tailoring or additions to the information provided therein. The last section, Section 8, proves that:

- Security objectives stated in Section 4 uphold all assumptions, counter all threats, and enforce all security policies identified in Section 3.
- Requirements specified in Section 5 implement all security objectives stated in Section 4.
- The design solution presented in Section 6 implements all requirements stated in Section 5.
- Claims made about PP compliance and achievement of the specified strength of function (SOF) are valid.

This proof is derived from a correlation analysis and consistency and completeness checks among the cited sections. Exhibit 3 summarizes the interaction among the sections of an ST.

As a guide, STs range from 100 to 200 pages in length, with the average distribution of pages per section as follows:

- Section 1. Introduction, 5 percent
- Section 2. TOE Description, 10 percent
- Section 3. TOE Security Environment, 10 percent
- Section 4. Security Objectives, 10 percent
- Section 5. Security Requirements, 20 percent
- Section 6. TOE Summary Specification, 20 percent
- Section 7. PP Claims, 5 percent
- Section 8. Rationale, 20 percent

It is important to understand the relationship between a PP and an ST. Sections 6 and 7, the TOE Summary Specification and PP Claims, are unique to an ST. In contrast, the first five sections of an ST mirror a PP. If an ST claims complete compliance with a PP, Sections 3 to 5 of the corresponding PP may be referenced rather than copied. However, the usual practice is to provide additional implementation-specific details in these sections. Section 8, Rationale, is similar to a PP Rationale; it includes two additional subsections: the TOE Summary Specification and the PP Claims Rationales. Exhibit 4 explains the relationship between sections of an ST and those of a PP.

A separate ST is written for each component TOE, should the referenced PP reflect a composite TOE. Some assumptions, threats, organizational security policies, security objectives, SFRs, and SARs contained in the PP will apply to all component TOEs; others will apply to only one component TOE. For example, element operations may be performed differently for each component TOE. In either case, the applicable assumptions, threats, organizational security policies (OSPs), security objectives, SFRs, and SARs

Exhibit 3. Interaction Among Section of an ST

ST Section	Purpose	Source
1. Introduction	identify nature, scope, and status of an ST	
2. TOE Description	describe system type, architecture, logical and physical security boundaries	
3. TOE Security Environment	state assumptions, identify threats, and cite security policies applicable to the TSF	
4. Security Objectives	delineate security objectives for the TOE and the operational environment	derived from an analysis of the assumptions, threats, and security policies articulated in Section 3
5. Security Requirements	implement security objectives through a combination of SFRs and SARs	derived from an analysis of the security architecture and boundaries (Section 2) and the perceived risk of compromise (Section 3)
6. TOE Summary Specification	deploy functional security mechanisms and security assurance measures in response to SFRs and SARs	derived from an analysis of security requirements in Section 5, current technology, and industry best practices
7. PP Claims	identify applicable PP and any tailoring and additions that were made	derived from a comparison of Sections 2 to 5 of the PP and ST
8. Rationale	demonstrate/prove that: (a) Section 4 security objectives are responsive to the Section 3 TOE security environment, (b) Section 5 requirements implement all Section 4 security objectives (c) Section 6 design solution implements all Section 5 requirements, and (d) PP compliance and SOF claims are valid	derived from a correlation analysis, consistency, and completeness checks among Sections 2–7

are incorporated into the ST for each component TOE. Exhibit 5 illustrates this concept.

4.2 Section 1: Introduction

The first section of an ST, Introduction, is divided into two subsections: ST Identification and ST Overview.

4.2.1 ST Identification

Information provided in the Identification subsection is used to properly catalog, index, and cross-reference an ST in registries maintained by the local National Evaluation Authority and other Common Criteria Recognition Agreement (CCRA) participants. The first field in the Identification section is the ST name; the second is the ST identifier. The first field simply states the ST name,

Exhibit 4. Similarities and Differences between Sections in a PP and Sections in an ST

PP Section	Interaction	ST Section
1. Introduction	• PP and ST essentially identical • SOF added to ST Overview	1. Introduction
2. TOE Description	• ST focus is on security architecture, logical and physical security boundaries. • PP focus is on system owner assets and TOE boundaries	2. TOE Description
3. TOE Security Environment	• PP and ST essentially identical • ST adds distinction between threats addressed by TOE and those addressed by operational environment	3. TOE Security Environment
4. Security Objectives	• PP and ST are essentially identical	4. Security Objectives
5. Security Requirements	• PP and ST are essentially identical • ST may complete element operations not performed in PP • ST may resolve dependencies not resolved in PP • ST may decompose SFRs into a lower level of detail	5. Security Requirements
6. PP Application Notes	N/A—(no corresponding section in an ST)	
	N/A—(no corresponding section in a PP)	6. TOE Summary Specification
	N/A—(no corresponding section in a PP)	7. PP Claims
7. PP Rationale	• first two subsections in PP and ST are essentially identical • ST adds TOE Summary Specification and PP Claims rationales	8. ST Rationale

while the second includes the version and date of the ST. The third field lists keywords associated with the ST, such as technology type, product category, development or user organization, and brand names. The fourth field cites the EAL to which a conformant TOE will be evaluated. The fifth and last field of the Identification indicates the current evaluation status of the ST. Exhibit 6 presents two examples of an ST Identification subsection.

4.2.2 ST Overview

The ST Overview provides a brief description of the ST and sets the context for the rest of the document. The ST Overview consists of five fields. The first field, which is also called ST Overview, is a stand-alone narrative that summarizes the security solution presented by the ST. It should be limited to a few paragraphs. The second field conveys the strength of function (SOF) requirement, which may be expressed as a minimum SOF for the entire ST or as an explicit SOF on a case-by-case basis. The third field lists PPs and STs that are related to this one and any documents referenced in the ST. Related STs could include other component STs that are part of the same composite TOE. Referenced documents may include organizational security

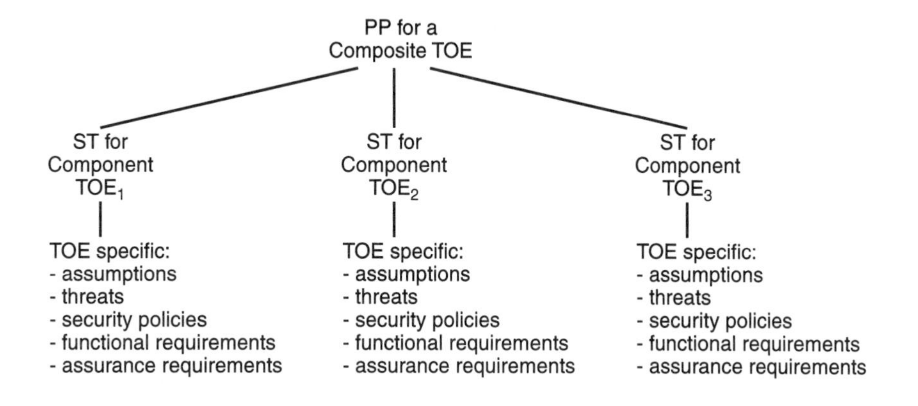

Exhibit 5. Relationship between an ST and a PP for a Composite TOE

standards and policies, national laws and regulations, and CC publications. The fourth field, ST Organization, explains the content and structure of the ST. This is the only "boilerplate" field in an ST. The fifth field defines acronyms as they are used in the ST.

4.3 Section 2: TOE Description

Section 2 of a PP contains four key pieces of information that serve as inputs to the developer:

1. General description of TOE functionality
2. List of TOE asset types and sensitivities
3. Access control rights and privilege for these assets
4. Definition of TOE boundaries

Exhibit 6. ST Identification Examples

Example 1: Composite TOE for a System

1.1 Security Target Identification
- 1.1.1 ST Name: High Assurance Remote Access
- 1.1.2 ST Identifier: U.S. DoD Remote Access PP for High Assurance Environments, version 1.0, May 2000
- 1.1.3 Keywords: remote access, network security, remote unit, communications server
- 1.1.4 EAL: EAL 5
- 1.1.5 ST Evaluation Status: Formal evaluation by a CCTL TBD.

Example 2: Monolithic TOE for a CCOTS Product

1.1 Security Target Identification
- 1.1.1 ST Name: Medium Assurance Traffic-Filter Firewall
- 1.1.2 ST Identifier: U.S. DoD Traffic-Filter Firewall PP for Medium Robustness Environments, version 1.0, January 2000.
- 1.1.3 Keywords: information flow control, firewall, packet filter, network security
- 1.1.4 EAL: EAL 2 augmented
- 1.1.5 ST Evaluation Status: Informal evaluation complete, formal evaluation by a CCTL TBD.

A developer may simply restate this information; however, it is preferable to provide some value-added analysis, especially if the ST is for a component TOE or a system. Restating information tells the customer and evaluator that the information has been read. In contrast, performing value-added analyses communicates that the information has been evaluated and is understood. The TOE Description consists of three subsections: System Type, Architecture, and Security Boundaries.

4.3.1 System Type

The System Type subsection introduces the TOE by defining the type of technology it represents. For a monolithic TOE or commercial "off the shelf" (COTS) product this is straightforward: application-level firewall, traffic-filter firewall, intrusion detection system (IDS) sensor, database access control utility, and so forth. More detail is provided for a system or composite TOE and the constituent component TOEs, such as the types of technology in each component TOE and what that technology performs within the framework of the overall composite TOE. The intended use of the TOE is defined at this point, as well as any dependencies on external IT entities. Some of this information may be derived from an analysis of the general functionality description in the PP and synthesis of TOE implementation specifics. The System Type subsection is generally no more than a couple paragraphs long. Exhibit 7 presents two sample System Type definitions: one for a backbone wide area network (WAN) and the other for a COTS product. Two statements contained in the first example are noteworthy because they help the reader envision the context in which the TOE will operate now and in the future: (1) the mention of a planned IP expansion, and (2) the fact that no traditional human users are included.

4.3.2 Architecture

The second subsection in the TOE Description presents a high-level view of the TOE architecture by expanding upon the System Type information. Major hardware, firmware, and software (system and application) platforms, components, and modules are cited, along with any hierarchical or structural dependencies or interactions. Specific version numbers and configurations are listed. Supplemental charts and diagrams illustrate these relationships.

A main objective of this subsection is to differentiate the TOE and TOE security architectures. The TOE architecture defines the context in which the TOE security function (TSF) must operate. The TSF consists of all TOE hardware, software, and firmware required for the correct enforcement of the TOE security policy.[19] The TOE security architecture defines an implementation-specific instance of the TSF within the constraints imposed by the TOE architecture and the asset types and sensitivities defined in the PP. If the ST is for a security-specific COTS product, the TOE and security architectures

Exhibit 7. ST System Type

Example 1: System

2.1 System Type

ABC is a nation-wide telecommunications backbone network which operates at ISO/OSI Reference Model layers 1–3. The common infrastructure consists of a combination of a fiber optic cable plant, including SONET rings, and an intelligent ATM and Frame Relay switching fabric. A limited IP capability is provided as well, with more planned for the future. High speed transmissions are supported, such as OC3, T3, and T1. Reliability, maintainability, and availability (RMA) and latency requirements dictate the use of redundant equipment, diverse paths, and hot standbys. To accommodate multiple security levels, some equipment is part of the public switched network, while other equipment is dedicated. ABC does not have any human users or end users in the traditional sense; rather, it interfaces with local communications equipment.

Example 2: COTS Security Product

2.1 System Type

The XYZ Guard is a network security device that uses the National Security Agency's (NSA) Fortezza cards to provide multi-level secure (MLS) services to legacy networks, i.e. Internet Protocol (IP) networks that operate in System High mode. XYZ Guards protect enclaves or individual hosts. Within a network, XYZ Guard is in-line between the host and the network. XYZ Guards operate on standard IP datagrams. The XYZ Guard can also serve as a firewall or an in-line encryptor.

may be the same. In contrast, if the ST is for a component TOE that is part of a system, major distinctions between the TOE and security architectures may exist. Exhibit 8 contains a sample architecture description for a COTS security product.

4.3.3 Security Boundaries

The third subsection of the TOE description is Security Boundaries, the purpose of which is to define the logical and physical security boundaries of the TOE. This contrasts with the end of Section 2 in a PP, which defines TOE boundaries. An evaluator uses the information in this section to determine the scope and boundaries of the evaluated configuration; hence, an accurate and concise description of what is and is not included within the boundaries is warranted.

The physical security boundary represents the perimeter to which TOE security functions are effective. IT entities within this perimeter are under the TSF scope of control (TSC) and are protected to the extent and in the manner specified. IT entities outside this perimeter are not within the TSC and may or may not be trusted. The TSC represents the set of interactions that can occur with or within a TOE and are subject to the rules of the TOE security policy.[22,117] The physical security boundary is defined by stating exactly which hardware, firmware, and software platforms, components, and modules comprise the implementation-dependent instantiation of the TSF. Version numbers, configuration options, interfaces, system utilities, initialization parameters, application modules, and so forth are identified. In addition, the TSF interface (TSFI) is described — the set of interfaces (man–machine or application program) through which resources are accessed or information is obtained

Exhibit 8. ST Architecture Example

Example: COTS Product

2.2 Architecture

The XYZ Guard is an enclosed unit containing a 486 motherboard and two Ethernet processors. The unit has two PCMCIA card slots, two Ethernet ports, and a serial port. One PCMCIA slot is used by a card which contains a digitally signed version of the Guard software. The second PCMCIA slot is used by a User Fortezza card that performs hashing, digital signature, key generation, and encryption functions. The User Fortezza card hosts eight digital certificates: user, configuration, audit, certification revocation, routing, local authority, root, and root authority.

XYZ Guards provide secure logical separation between networked domains (PCs, workstations, servers that are networked together and operate at the same security level). An administration system defines the security level and network configuration information for each domain and stores this in the User Fortezza card. A Local Authority Fortezza card is required to generate User Fortezza cards. Configuration information is verified by checking the output of the serial port during initial setup.

from the TSF.[20] In essence, the physical security boundary definition adds an extra level of detail to the information provided in the architecture subsection.

The logical security boundary is defined in terms of security services or features. A common convention is to: (1) discuss what security services are provided by each of the items identified as being within the physical security boundary, and (2) group this information under headings the same as the names of functional classes, such as Security Audit, User Data Protection, Identification and Authentication, Security Management, and so forth.

The final statement in this subsection clarifies what is not within the TOE security boundaries. A standard phrase, quoted below, is followed by a list of specific items that are excluded:

> Software and hardware features outside the scope of the defined TSF and thus not evaluated are: [list supplied by developer].

Exhibit 9 illustrates a partial definition of physical and logical security boundaries.

4.4 Section 3: Security Environment

The third required section of an ST, the TOE Security Environment, defines the nature and scope of TOE security. The ST Security Environment mirrors the PP Security Environment to the extent that this is appropriate; differences between the two result from the fact that a PP is implementation independent while an ST is implementation dependent. The Security Environment consists of three subsections: Assumptions, Threats, and Organizational Security Policies.

4.4.1 Assumptions

The Assumptions subsection in a PP relays pertinent domain knowledge to developers to help them understand the overall framework of the TOE. In an

Exhibit 9. TOE Security Boundary Definitions

Example 1: Physical Security Boundaries

2.3 Security Boundaries

2.3.1 Physical Boundaries

The items described in Table 1 are within the physical security boundary of the TOE.

Table 1: Identification of TOE Physical Security Boundaries

TOE	Component/Module
Workstation	• AMD Athlon 1-GHz processor • 512 MB RAM • 20-GB hard disk • 3.5-inch floppy drive • CD-RW drive • keyboard • mouse • serial port • 2 USB ports • 15 inch, high resolution flat panel color monitor • power cord • 10/100-Mbps Ethernet interface • Microsoft Windows Me Professional • file system • security subsystem • event log services • registry services • Corel Office 2002

Example 2: Logical Security Boundaries

2.3. Security Boundaries

2.3.2. Logical Boundaries

2.3.2.x. *User Data Protection*

The FTP and Telenet security servers provide authentication and protection from malformed service requests, while the HTTP and SMTP security servers provide application level protection. Module A ensures that information contained in packets from previous sessions is no longer accessible once the session has been completed. The storage and processing of data packets is managed to ensure that no residual information is transferred to future sessions. Module B performs the inspection process by applying the TSP rules. Module C renders a decision on whether each packet should be accepted, rejected, or dropped.

ST, the developer confirms PP assumptions (or a variant of them) to the customer and evaluator. Assumptions are made about the intended use, operational environment, and connectivity of the TOE, as well as roles and responsibilities. Any environmental constraints or operational limitations are clarified. If the ST is for a component TOE, only the PP assumptions that pertain to that TOE must be evaluated. Like a PP, ST assumptions cannot be used to mitigate threats. The developer has four options concerning assumptions:

1. PP assumptions may simply be restated verbatim.
2. PP assumptions may be modified or tailored to reflect TOE implementation specifics.

3. New assumptions may be added.
4. PP assumptions may not be carried forward if they are inappropriate or not applicable.

Exhibit 10 presents sample ST assumptions, using Exhibit 13 from Chapter 3 as input.

4.4.2 Threats

Section 3.2 of a PP identifies potential threats posed by TOEs and determines their risk mitigation priority as a function of the likelihood of occurrence and severity of consequences. A developer may simply restate this information; however, it is preferable to perform some value-added analysis, especially if the ST is for a component TOE or a system.

The first table in Section 3.2 of a PP itemizes potential threats in a hierarchical fashion. This table serves as an input to the developer, who analyzes it to determine which threats are:

- Addressed by the TOE design solution defined in the ST
- Addressed by the TOE security environment defined in the ST
- Not applicable to either the TOE or TOE environment

Each threat identified in the PP must fall into one of these three categories. The rationale for assigning threats to either of the first two categories is explained in Section 8.3 of the ST. If a threat falls into the third category, this is explained in Section 8.1 of the ST. In addition, the developer may identify

Exhibit 10. ST Assumptions

3.1 Assumptions

This subsection restates assumptions that were made in the PP and tailors them, when appropriate, to reflect TOE implementation specifics. Four types of assumptions are expressed: intended use, operational environment, connectivity, and personnel roles and responsibilities.

3.1.1 Intended Use

- A1* TOE components rely on the SUN Solaris operating system and utilities that are assumed to be installed and operated in a secure manner and in accordance with the ST and other relevant documentation and procedures.
- A2* All TOE files are assumed to be protected from unauthorized access by the SUN Solaris operating system.
- A3 The TOE, including the TSF, will meet specified RMA requirements.
- A4* The TOE only consists of the assets described in Section 2 of the referenced PP and only processes data of the sensitivities indicated therein.
- A5 Audit information is reviewed and analyzed on a periodic basis in accordance with the network security policy.
- A6 Cryptographic methods will be resistant to cryptanalytic attacks and be of adequate robustness to protect sensitive data.

* tailored assumption.

new potential threats not listed in the PP. Exhibit 11 illustrates this process, using Exhibit 14 from Chapter 3 as input.

The second table in Section 3.2 of a PP assigns a risk mitigation priority for each threat. This table serves as an input to the developer, who analyzes it to determine the residual risk of each threat *after* the countermeasures contained in the design solution (technical, operational, and procedural) have been deployed. In essence, this is an assessment of the robustness and resilience of the proposed design solution. At this point, only the assessment is made; the justification for this assessment is provided in Section 8.3 of the ST. Exhibit 12 illustrates this process, using Exhibit 15 from Chapter 3 as input.

4.4.3 Organizational Security Policies

Section 3.3 of a PP cites organizational security policies (OSPs) that are relevant to the TOE or the TOE environment. OSPs include rules, procedures, and practices that an organization imposes on an IT system to protect its assets.[19] Local, national, or international laws and regulations may impose additional OSPs — for example, privacy requirements. To ensure that the correct version of such laws, regulations, and so forth are being adhered to, it is useful to cite the source of OSPs.

Organizational security policies are unique to each organization and its mission and assets; the customer is the owner of OSPs. As a result, the developer is somewhat limited in terms of what actions can be performed with PP OSPs:

- OSPs can be restated verbatim.
- OSPs can be tailored to reflect TOE implementation specifics, as long as the original intent is not changed.

Organizational security policies cannot be added or deleted. To enhance clarity, OSPs may be assigned to the TOE or the TOE security environment. If the ST is for a component TOE, only those OSPs that are applicable to that TOE must be articulated.

4.5 Section 4: Security Objectives

Customers state their security objectives in PPs and divide the responsibility for them among the TOE, the IT environment, and the non-IT environment. In addition, the objectives are categorized as being preventive, detective, or corrective. The customer is the owner of the security objectives. As a result, a developer responds to a customer's security objectives. Blindly restating the

Exhibit 11. ST Threat Identification

Table x Potential Threats Addressed by the TOE or the TOE Environment

#	Threat	TOE	TOE Environment
T1	An undetected compromise of assets may occur as a result of:		
	an authorized user performing actions the individual is not authorized to perform	X	
T1b	an attacker (insider or outsider) masquerading as an authorized user and attempting to perform actions that individual is authorized to perform	X	
T1c	an attacker (insider or outsider) gaining unauthorized access to information or resources by impersonating an authorized user.	X	X
T1d	an authorized or unauthorized user accidentally or intentionally blocking staff access to TOE devices	X	X
T1e	an unauthorized user gaining control of the TOE	X	X
T1f	an unauthorized user rendering the TOE inoperable	X	X
T1g	an unauthorized person attempting to bypass security	X	
T1h	an unauthorized person repeatedly trying to guess identification and authentication data	X	
T1i	an unauthorized person using valid identification and authentication data fraudulently	X	
T1j	an unauthorized person or external IT entity viewing, modifying, or deleting security-relevant information transmitted to a remote authorized user or administrator	X	
T2	An authorized user may access information or resources without having permission from the person who owns or is responsible for the information or resource	X	X
T3	An attacker may eavesdrop on or otherwise capture data being transmitted across a network:		
T3a	unauthorized users performing traffic analysis		X
T3b	an authorized or unauthorized user using residual information from previous information flows	X	
T4	An authorized user or unauthorized outsider consumes global resources in a way that compromises the ability of other authorized users to access or use those resources:		
T4a	circuit jamming (voice or data)		X
T4b	DoS and DDoS attacks (voice or data)	X	X
T4c	theft of service		X
T5	A user may intentionally or accidentally transmit sensitive information to users who are not cleared to see it	X	
T6	A user may participate in the transfer of information either as originator or recipient and then subsequently deny having done so.	X	
T7	An authorized user may export information in soft- or hard-copy form, which the recipient subsequently handles in a manner that is inconsistent with its sensitivity designation.	X	X
T8	The integrity and availability of information may be compromised due to:		
T8a	user errors, firmware errors, hardware errors, or transmission errors	X	X
T8b	the unauthorized modification or destruction of the information by an attacker	X	X

Exhibit 11. ST Threat Identification (continued)

Table x Potential Threats Addressed by the TOE or the TOE Environment

#	Threat	TOE	TOE Environment
T8c	human errors or a failure of software, firmware, hardware, or power supplies which causes an abrupt interruption to operations, resulting in the loss or corruption of critical data	X	X
T8d	aging of storage media or improper storage or handling of storage media		X
T8e	an authorized user unwittingly introducing a virus into the system	X	X
T8f	an authorized user may introduce unauthorized software into the system		X
T8g	an authorized or unauthorized user inserting malicious code or backdoors	X	X
T8h	an unauthorized person reading, modifying, or destroying security critical configuration information	X	
T8i	failure to perform adequate system backups		X
T8j	accidental or intentional deletion	X	X
T8k	insertion of bogus data	X	X
T8l	unauthorized modification of data (payload or header)	X	X
T9	An attacker could observe the legitimate use of a resource or service by a user, when the user wishes their use of that resource or service to be kept confidential	X	X
T10	An authorized user may intentionally or accidentally observe stored information that the user is not cleared to see		X
T11	Security-critical components may be subject to physical attack or operational environmental failures, which may compromise security		X
T12	An authorized insider or unauthorized outsider may accidentally or intentionally cause security-relevant events not to be recorded or traceable:		
T12a	legitimate audit records may be lost or overwritten	X	
T12b	audit records may not be attributed to time of occurrence	X	
T12c	audit records may not be attributed to actual source of activity	X	
T12d	people not being accountable for their actions because audit records are not reviewed	X	
T12e	compromises of user or system resources may go undetected for long periods of time	X	X
T13	Weaknesses in the architecture, design, implementation, operation, or maintenance may precipitate security failures or compromises	X	X
T14	An authorized insider or unauthorized outsider may cause the improper restart and/or recovery from failure of hardware, software, or firmware that causes a security compromise	X	X
T15	Changes in operational environment may introduce or exacerbate vulnerabilities		X
T16	A knowledgeable adversary may circumvent unexpected limitations or latent defects in countermeasures and mitigation strategies	X	X
T17	The definition, implementation, and enforcement of access control rights and privileges may be done in a manner that undermines security.	X	
T18	Natural disasters or acts of war or terrorism could result in critical operations being interrupted or halted.		X

Exhibit 11. ST Threat Identification (continued)

Table x Potential Threats Addressed by the TOE or the TOE Environment

#	Threat	TOE	TOE Environment
T19	Compromise of assets may occur as a result of actions taken by careless, willfully negligent or hostile administrators or other privileged users:		
T19a	improper operation of hardware, software, and or firmware	X	X
T19b	premature hang-up of voice circuit		X
T19c	premature shut-down of PVC or VPN		X
T19d	OPSEC procedures being inadequate		X
T19e	OPSEC procedures being poorly written		X
T19f	users and administrators unfamiliar with OPSEC procedures		X

customer's security objectives in an ST is not recommended. Instead, the developer should thoroughly analyze each objective to determine:

- Is each objective still valid, given the developer's response to the security environment described in Section 3 of the ST? If yes, the objective should be carried forward to the ST verbatim. If no, should the objective be (a) deleted, (b) modified, (c) reassigned, or (d) reclassified?
- Should any new objectives be added?

In general, the majority of PP security objectives can usually be carried forward in the ST. A certain number of them remain, however, that require some sort of rework to be consistent with the ST. A few may be deleted if they are no longer applicable or are incompatible with the proposed security solution. Some may be reworded to incorporate specifics of the design solution. Others may have to be recategorized as being applicable to the TOE, the IT environment, or the non-IT environment, or reclassified as being preventive, detective, or corrective. It is also possible that some new security objectives may have to be added, particularly if any assumptions or threats were added or tailored in Section 2 of the ST.

4.6 Section 5: Security Requirements

The fifth required section of an ST is Security Requirements. Security requirements are described in four subsections: Security Functional Requirements, Security Assurance Requirements, Security requirements for the IT Environment, and Security Requirements for the Non-IT Environment.

4.6.1 Security Functional Requirements (SFRs)

The customer states their security functional requirements in subsection 5.1 of a PP and, in the case of a composite TOE, identifies to which component

Exhibit 12. ST Threat Assessment

Table x: Assessment of Residual Risk

#	Threat	Severity of Consequences (note 1)	Likelihoodof Occurrence (note 2)	Risk Mitigation Priority	Residual Risk
T1	An undetected compromise of assets may occur as a result of:				
T1a	an authorized user performing actions the individual is not authorized to perform	marginal to critical	occasional	high	low
T1b	an attacker (insider or outsider) masquerading as an authorized user and attempting to perform actions that individual is authorized to perform	marginal to critical	occasional	high	low
T1c	an attacker (insider or outsider) gaining unauthorized access to information or resources by impersonating authorized user	marginal to critical	occasional	high	low
T1d	an authorized or unauthorized user accidentally or intentionally blocking staff access to TOE devices	marginal to critical	occasional	high	low
T1e	an unauthorized user gaining control of the TOE	marginal to critical	remote	medium to high	low
T1f	an unauthorized user rendering the TOE inoperable	marginal to critical	remote	medium to high	low
T1g	an unauthorized person attempting to bypass security	marginal to critical	frequent	medium to high	low
T1h	an unauthorized person repeatedly trying to guess identification and authentication data	marginal to critical	frequent	medium to high	medium
T1i	an unauthorized person using valid identification and authentication data fraudulently	marginal to critical	probable	medium to high	low
T1j	an unauthorized person or external IT entity viewing, modifying, and/or deleting security-relevant information transmitted to a remote authorized user or administrator	marginal to critical	occasional	medium to high	low
T2	An authorized user may access information or resources without having permission from the person who owns or is responsible for the information or resource	marginal to critical	remote	medium	low
T3	an attacker may eavesdrop on or otherwise capture data being transmitted across a network:				
T3a	unauthorized users performing traffic analysis	marginal	remote	low	low
T3b	an authorized or unauthorized user using residual information from previous information flows	marginal	remote	low	low
T4	An authorized user or unauthorized outsider consumes global resources in a way that compromises the ability of other authorized users to access or use those resources:				

Exhibit 12. ST Threat Assessment (continued)

Table x*: Assessment of Residual Risk*

#	Threat	Severity of Consequences (note 1)	Likelihood of Occurrence (note 2)	Risk Mitigation Priority	Residual Risk
T4a	circuit jamming (voice or data)	marginal to catastrophic	remote	high	low
T4b	DoS and DDoS attacks (voice or data)	marginal to catastrophic	remote	high	low
T4c	theft of service	marginal to catastrophic	remote	high	low
T5	A user may intentionally or accidentally transmit sensitive information to users who are not cleared to see it.	marginal to critical	remote	medium	low
T6	A user may participate in the transfer of information either as originator or recipient and then subsequently deny having done so.	marginal	remote	low	low
T7	An authorized user may export information in soft- or hard-copy form, which the recipient subsequently handles in a manner that is inconsistent with its sensitivity designation.	marginal to critical	occasional	high	low
T8	The integrity and availability of information may be compromised due to:				
T8a	user errors, firmware errors, hardware errors, or transmission errors	marginal to catastrophic	occasional	high	low
T8b	the unauthorized modification or destruction of the information by an attacker	marginal to catastrophic	remote	medium	low
T8c	human errors or a failure of software, firmware, hardware or power supplies which causes an abrupt interruption to operations, resulting in the loss or corruption of critical data	marginal to catastrophic	remote	medium	low
T8d	aging of storage media or improper storage or handling of storage media	marginal to catastrophic	remote	medium	low
T8e	an authorized user unwittingly introducing a virus into the system	marginal to catastrophic	frequent	high	medium
T8f	an authorized user may introduce unauthorized software into the system	marginal to catastrophic	frequent	high	medium
T8g	an authorized or unauthorized user inserting malicious code or backdoors	marginal to catastrophic	occasional	medium	low
T8h	an unauthorized person reading, modifying, or destroying security-critical configuration information	marginal to catastrophic	occasional	medium to high	low
T8i	failure to perform adequate system backups	marginal	occasional	medium	low
T8j	accidental or intentional deletion	marginal to critical	occasional	medium to high	low

Exhibit 12. ST Threat Assessment (continued)

Table x: Assessment of Residual Risk

#	Threat	Severity of Consequences (note 1)	Likelihood of Occurrence (note 2)	Risk Mitigation Priority	Residual Risk
T8k	insertion of bogus data	marginal to critical	occasional	medium to high	low
T8l	unauthorized modification of data (payload or header)	marginal to critical	occasional	medium to high	low
T9	an attacker could observe the legitimate use of a resource or service by a user, when the user wishes their use of that resource or service to be kept confidential	marginal to critical	occasional	high	low
T10	An authorized user may intentionally or accidentally observe stored information that the user is not cleared to see	marginal to critical	occasional	medium	low
T11	Security-critical components may be subject to physical attack and/or operational environmental failures, which may compromise security	insignificant to catastrophic	improbable	low	low
T12	An authorized insider or unauthorized outsider may accidentally or intentionally cause security-relevant events not to be recorded or traceable:	marginal to catastrophic	remote	medium	
T12a	legitimate audit records being lost or overwritten	marginal to catastrophic	remote	medium	low
T12b	audit records not being attributed to time of occurrence	marginal to catastrophic	remote	medium	low
T12c	audit records may not be attributed to actual source of activity	marginal to catastrophic	remote	medium	low
T12d	people not being accountable for their actions because audit records are not reviewed	marginal to catastrophic	remote	medium	low
T12e	compromises of user or system resources going undetected for long periods of time	marginal to catastrophic	remote	medium	low
T13	Weaknesses in the architecture, design, implementation, operation, or maintenance may precipitate security failures or compromises	marginal to critical	remote	medium	low
T14	An authorized insider or unauthorized outsider may cause the improper restart aor recovery from failure of hardware, software, or firmware that causes a security compromise	marginal to critical	remote	medium	low
T15	Changes in operational environment may introduce or exacerbate vulnerabilities	marginal to critical	remote	low	low
T16	A knowledgeable adversary may circumvent unexpected limitations or latent defects in countermeasures and mitigation strategies.	marginal to critical	remote	medium	low

Exhibit 12. ST Threat Assessment (continued)

Table x: Assessment of Residual Risk

#	Threat	Severity of Consequences (note 1)	Likelihoodof Occurrence (note 2)	Risk Mitigation Priority	Residual Risk
T17	The definition, implementation, and enforcement of access control rights and privileges may be done in a manner that undermines security	marginal to critical	remote	medium	low
T18	Natural disasters or acts of war or terrorism could result in critical operations being interrupted and/ or halted	marginal to catastrophic	improbable	low	low
T19	Compromise of assets may occur as a result of actions taken by careless, willfully negligent or hostile administrators or other privileged users:				
T19a	improper operation of hardware, software, or firmware	marginal to catastrophic	remote	medium	medium
T19b	premature hang-up of voice circuit	marginal to catastrophic	remote	medium	low
T19c	premature shut-down of PVC or VPN	marginal to catastrophic	remote	medium	low
T19d	OPSEC procedures being inadequate	marginal to catastrophic	remote	medium	low
T19e	OPSEC procedures being poorly written	marginal to catastrophic	remote	medium	low
T19f	users and administrators unfamiliar with OPSEC procedures	marginal to catastrophic	remote	medium	medium

Note 1: Standard severity definitions from IEC 61508 are used:

- *catastrophic*—loss of one or more major systems which may or may not be accompanied by fatalities or multiple severe injuries.
- *critical*—loss of a major system which may or may not be accompanied by a single fatality or severe injury.
- *marginal*—severe system damage which may or may not be accompanied by minor injuries.
- *insignificant*, system damage which may or may not be accompanied by single minor injury.

Note 2: Standard likelihood definitions from IEC 61508 are used:

- *frequent*—likely to occur frequently, 10^{-2}
- *probable*—will occur several times, 10^{-3}
- *occasional*—likely to occur several times over the life of a system, 10^{-4}
- *remote*— likely to occur at some time during the life of a system, 10^{-5}
- *improbable*—unlikely but possible to occur during the life of a system, 10^{-6}
- *incredible*— extremely unlikely to occur during the life of a system, 10^{-7}

TOEs they apply. The developer responds to these requirements in Subsection 5.1 of an ST. The developer may:

- Simply restate the requirements verbatim (selecting the applicable SFRs if the ST is for a component TOE).
- Complete element operations that were not performed in the PP.
- Resolve component dependencies that were not resolved in the PP.
- Refine or iterate functional components to reflect the proposed security solution.
- Specify audit requirements not supplied in the PP.

1. TOE Security Functions

SFR 1, SFR 2, SFR 3 → Security Function 1

SFR 10, SFR 11, SFR 12 → Security Function 2

Security Function 1, Security Function 2 → Security Mechanism 1

2. Assurance Measures

SAR 1, SAR 2, SAR 3 → Assurance Measure 1

SAR 10, SAR 11, SAR 12 → Assurance Measure 2

Assurance Measure 1, Assurance Measure 2 → Assurance Package 1

Exhibit 13. TOE Summary Specification Mapping

- Add new SFRs that are necessary because of the implementation-dependent nature of the ST.
- Reassign SFRs to security requirements for the IT or non-IT environment.
- Omit SFRs that are unnecessary, redundant, or conflicting.

The last seven bullets give the developer an opportunity to further decompose the requirements and add implementation-specific details.

4.6.2 Security Assurance Requirements (SARs)

Security assurance requirements define the criteria for evaluating PPs, STs, and TOEs and the security assurance responsibilities of developers and evaluators. The customer determines the appropriate EAL and hence the SARs. The developer simply restates the SARs, including the developer action items, content and presentation of evidence criteria, and evaluator action items, to indicate that they understand the nature and extent of security assurance activities to be performed. (If, for some reason, the developer concludes that the specified EAL is unattainable, excessive, or too weak, that discussion is captured in subsection 8.2 of the ST.)

4.6.3 Security Requirements for the IT Environment

In subsection 5.3 of a PP, the customer states their security requirements for the IT environment. They are derived from the operational environment in

Exhibit 14. TSF Mapping Example: Step 1

Functional Package	Long Name	SFR from PP	Long Name
GRD_ADM	Security administration	FMT_SMR.1	Security management roles
		FMT_MOF.1	Management of functions in TSF
		FIA_ATD.1	User attribute definition
GRD_INA	Identification and authentication	FIA_UID.2	User identification
		FIA_UAU.1	User authentication
GRD_FLO	Information flow control	FDP_IFC.1	Information flow control policy
		FDP_IFF.1	Information flow control functions
		FDP_RIP.2	Residual information protection
		FPT_RVM.1	Reference mediation
GRD_DFL	Default configuration	FMT_MSA.3	Management of security attributes
GRD_SEP	Isolation	FPT_SEP.1	Domain separation
GRD_AUD	Security Audit	FPT_STM.1	Time stamps
		FAU_GEN.1	Security audit data generation
		FAU_SAR.1	Security audit review
		FAU_SAR.3	Security audit review
		FAU_STG.1	Security audit event storage
		FAU_STG.4	Security audit event storage
NONE	NONE	FIA_AFL.1	Authentication failures
		FIA_UAU.4	User authentication
		FCS_COP.1	Cryptographic operation

which the customer intends to deploy the TOE. The developer responds to these requirements and interprets them in view of the proposed security solution. Again, the developer may address implementation-specific details by:

- Simply restate the requirements verbatim (selecting the applicable SFRs if the ST is for a component TOE).
- Complete element operations that were not performed in the PP.
- Resolve component dependencies that were not resolved in the PP.
- Refine or iterate functional components to reflect the proposed security solution.
- Specify audit requirements not supplied in the PP.
- Add new SFRs that are necessary because of the implementation-dependent nature of the ST.
- Reassign SFRs to security requirements for the TOE or non-IT environment.
- Omit SFRs that are unnecessary, redundant, or conflicting.

Exhibit 15. TSF Structure Example: Step 2

```
1.  TSF
    1.1  Security Administration Package
         1.1.1  start-up, shut-down
         1.1.2  create, write, edit, delete, read information flow control rules
         1.1.3  create, write, edit, delete, read user attributes
         1.1.4  set date and time
         1.1.5  create, delete, read, archive audit trail datas
         1.1.6  create system backups
         1.1.7  Initiate system recovery
    1.2  Identification and Authentication Package
         1.2.1  operating system level
         1.2.2  application system level
         1.2.3  human
         1.2.4  internal IT entities and processes
         1.2.5  external IT entities
    1.3  Information Flow Control Package
         1.3.1  control access to external resources
         1.3.2  control access to internal resources
    1.4  Default Configuration Package
         1.4.1  block traffic during installation, generation, and start-up
         1.4.2  block transactions during installation, generation, and start-up
    1.5  Isolation Package
         1.5.1  maintain each session in a separate logical domain
    1.6  Security Audit Package
         1.6.1  generate audit data
         1.6.2  selectable audit data review
         1.6.3  protected audit storage
         1.6.4  prevent loss of audit data
```

4.6.4 Security Requirements for the Non-IT Environment

Likewise, in subsection 5.4 of a PP, customers state their security requirements for the non-IT environment which are derived from organizational operational security procedures. The developer responds to these requirements and interprets them in view of the proposed security solution. Most security requirements for the non-IT environment are stated as explicit requirements. Even so, the developer may address implementation-specific details by:

- Simply restating the requirements verbatim (selecting the applicable SFRs if the ST is for a component TOE)
- Adding new SFRs that are necessary because of the implementation-dependent nature of the ST
- Reassigning SFRs to security requirements for the TOE or non-IT environment
- Omiting SFRs that are unnecessary, redundant, or conflicting

4.7 Section 6: TOE Summary Specification

The sixth section, TOE Summary Specification (TSS), depicts the developer's design solution, which addresses the security requirements stated in the PP.[19] The TSS contains requirement traceability matrices that define:[19,22]

- Specific IT security functions that satisfy each SFR identified in Section 5
- Exact security mechanisms or techniques that are used to implement each IT security function
- Precise security assurance measures that are employed to satisfy each SAR identified in Section 5

This mapping is performed to ensure that: (1) all requirements (SFRs and SARs) are accounted for by the design solution, and (2) no new unspecified functionality has been introduced. Each SFR must map to at least one function; likewise, each function must map to at least one SFR. A security function may correspond to a functional package, depending on the organization of the PP or ST. Each SAR must map to at least one security assurance measure, while each security assurance measure must map to at least one SAR. Security functions are then mapped to specific security mechanisms and security assurance measures are mapped to a specific assurance package (see Exhibit 13). This section consists of two subsections: TOE Security Functions and Security Assurance Measures. Information presented in the first subsection may be identical to that generated for the ADV_FSP security assurance activities.

4.7.1 TOE Security Functions

This subsection of an ST, TOE Security Functions (TSF), describes in detail the security functions performed by the TOE and the specific security mechanisms that implement them. The TSF consists of all TOE hardware, software, and firmware that enforce the TOE security policy.[19] This information is organized by functional packages, classes, and families to aid readability and understandability. Clear distinctions are made between the TOE and the TSF, consistent with the logical and physical security boundaries defined in Section 2 of the ST. However, if the TOE is a COTS product that only performs security functions, the TOE and the TSF may be identical. In addition, the TSF Scope of Control (TSC) and TOE Security Function Interfaces (TSFI) are clearly delineated. The TSC is the set of interactions that can occur with or within a TOE and that are subject to the rules defined in the TOE Security Policy.[20,117] The TSFI is the set of interfaces through which either resources are accessed or information is obtained from the TSF;20 both actions require TSF mediation. These interfaces can be interactive man–machine interfaces or application program interfaces.

A six-step process is followed to elucidate the TSF, as shown below. Each step in the process provides an extra level of implementation detail. All such information must meet the ASE security assurance evaluation criteria of completeness, consistency, coherence, correctness, and unambiguousness. Textual descriptions should be supplemented with tables and diagrams whenever possible:

Step 1 — TSF packages are mapped to SFRs.
Step 2 — The TSF structure is defined.
Step 3 — Specific security mechanisms that implement security functions are identified.
Step 4 — The capture of audit requirements is explained.
Step 5 — Satisfaction of SOF requirements is demonstrated.
Step 6 — The implementation of management requirements is described.

Exhibit 14 illustrates the first step. The six TSF packages are listed in the first two columns by their short and long names. The SFRs are mapped to the corresponding TSF packages. All SFRs must be mapped: those from the PP; those added or modified by the ST; and SFRs for the TOE, the IT environment, and the non-IT environment. As noted earlier, each SFR must map to at least one function; likewise, each function must map to at least one SFR. In this example all but three of the SFRs map to one and only one TSF package. FIA_AFL.1, FIA_UAU.4, and FCS_COP.1 do not map to any TSF package. Consequently, in order for this ST to be certified the developer must do three things:

1. In Section 6 of the ST, TOE Summary Specification, a statement must be made to explain why each of the three SFRs were not incorporated into the security solution.
2. In Section 7 of the ST, Protection Profile Claims, partial conformance with the referenced PP is claimed.
3. In Section 8 of the ST, Rationale, a justification is provided for exclusion of the three SFRs.

Exhibit 15 illustrates the second step. The hierarchical relationship between TSF packages is depicted and the major functions of each TSF package are listed. In this example, the TSF contains six functional packages: security administration, identification and authentication, information flow control, default configuration, isolation, and security audit. All six TSF packages are at the same level and the functions they perform are self-contained. The security administration package consists of seven functions, while the isolation package consists of only one function.

Exhibit 16 illustrates the third step. The security mechanisms that implement each security function are described. Default settings, constraints, operational parameters, and algorithmic details are described. In this example, the security mechanisms that implement the security audit TSF package are designated. (Note that this is a high-level example; in an actual ST, each 1.6.*x.x* entry would contain a few paragraphs to a few pages of detail.)

Exhibit 17 illustrates the fourth step. Details are provided about the level of audit requirements specified and the events that can be audited and captured for each applicable SFR.

Exhibit 16. Mapping Security Mechanisms to TSF Packages: Step 3

1.6 Security Audit Package

1.6.1 Generate audit data

1.6.1.1 Module A generates monitoring information, such as audit trails, security event logs, and alerts for SNMP traps.

1.6.1.2 Module B transmits audit trails, security event logs, and alerts generated by Module A to the Daemon component for further processing.

1.6.1.3 Audit records are generated for the following events:

- start-up and shut-down of the TOE and TSF,
- modifications to the group of users that are part of the authorized administrator role,
- all use of user identification mechanisms, including the user identity provided,
- all use of authentication mechanisms,
- all decisions on requests for information flow,
- create, write, delete, edit, and read information flow security policy rules that permit or deny information flows,
- create, write, delete, edit and read user attributes,
- set and modify system time and date,
- archive, create, delete, read, and purge the audit trail, and
- backup and recovery transactions.

1.6.2 Selectable audit data review

1.6.2.1 The XYZ tool has a GUI that permits authorized system administrators to read, search, and sort audit records based on event type, date and time, and range of IP addresses.

1.6.3 Protected audit storage

1.6.3.1 Audit files are protected by a secure file system (NTFS).

1.6.3.2 Only authorized users have access to audit files.

1.6.3.3 NTFS detects attempted modifications of audit files.

1.6.3.4 NTFS records successful and unsuccessful attempts to create, read, write, edit, delete, change permissions, and/or take ownership of audit files.

1.6.4 Prevent loss of audit data

1.6.4.1 TOE and TSF operation is halted if the system is unable to capture or record audit data.

1.6.4.2 Only authorized system administrators may perform functions until the above situation is corrected.

Exhibit 18 illustrates the fifth step. The criteria employed by the design solution to meet each strength of function requirement are cited. Minimum SOF requirements may be specified for an entire TOE, or explicit SOF requirements may be specified for each unique situation. The TSS provides a design solution in either situation. The TSS is limited to describing the design solution; the Rationale in Section 8 explains why the proposed solution is robust and resilient enough.

The sixth step is to describe the implementation of management requirements. ISO/IEC 15408-2 includes management requirements as part of the definition of the security management class (FMT) components. Management requirements are provided for a developer's consideration — management requirements can be selected from those contained in the standard and new requirements can be developed. For example, the following management requirements are listed for FMT_REV.1.[20]

Exhibit 17. Sample TTSS for Audit Requirements: Step 4

SFR	Audit Level	Auditable Event(s)
FMT_SMR.1	Minimal	• modifications to the group of users that are part of the authorized administrator role • identity of the authorized administrator performing the above modification • user identify being associated with the authorized administrator role
FIA_UID.2	Basic	• all use of the user identification mechanism • user identities provided to the toe security function (tsf)
FIA_UAU.1	Basic	• all use of the authentication mechanism • user credentials provided to the toe security function (tsf)
FIA_AFL.1	Minimal	• threshold reached for unsuccessful authentication attempts • subsequent restoration of user authentication ability by authorized administrator • identity of offending user • identity of authorized system administrator who performed the restoration
FDP_IFF.1	Basic	• all decisions made on requests for information flow • presumed addresses of the source and destination subjects
FPT_STM.1	Minimal	• changes to system time • identity of authorized system administrator who changed system time

Management: FMT_REV.1

The following actions could be considered for the management functions:

a) managing the group of roles that can invoke revocation of security attributes;
b) managing the lists of users, subjects, objects, or other resources for which revocation is possible;
c) managing the revocation rules.

In other words, the standard suggests managing (a) who can perform revocation functions, (b) what entities are subject to revocation, and (c) how the threshold for performing revocation will be determined. These three items correspond directly to the FMT_REV.1 element definitions:[20]

FMT_REV.1.1 *The TSF shall restrict the ability to revoke security attributes associated with the [selection: users, subjects, objects, other additional resources] within the TSC to [assignment: the authorized identified roles].*

FMT_REV.1.2 *The TSF shall enforce the rules [assignment: specification of revocation rules].*

The selection operation corresponds to item (b) of the management requirements, while the assignment operations correspond to items (a) and (c) of the management requirements.

Exhibit 18. Sample TSS Strength of Function Criteria: Step 5

Example 1: SOF–basic

All passwords used by the TOE or TSF will meet the following criteria:

- minimum of eight characters long
- valid characters limited to: a–z, A–Z, 0–9, !#$%&*
- must contain at least one capital letter and one numeric character
- must be changed every 3 months
- reuse of previous passwords cannot be reused within 36 months

Example 2: SOF–medium

The authentication mechanism for general users of TOE security services will meet the following criteria:

- secret key length of at least 64 characters
- MD5 hash algorithm
- random number generator complies with statistical random number generator tests and continuous random number tests specified in FIPS 140–2 for level 2

Example 3: SOF–medium

A strength of function level of SOF medium counters a medium attack level. This SOF requirement is met by using the cryptographic services provided by the User Fortezza Card.

The developer might want to add a fourth management requirement concerning revocation time frames:

d. Managing the frequency with which revocation is performed on a routine basis and how quickly revocation can be performed on an emergency basis

To do so, an explicit requirement would be defined:

Explicit requirement

FMT_REV.1.3 *The TSF shall perform revocation on demand and complete the transaction within 5 seconds.*

The security solution presented in the TSS describes how the management functions are implemented. For the example here, specifics would be provided about how firmware/software modules:

- Revoke security attributes and what attributes can be revoked
- Limit the capability to do this to authorized users
- Perform the revocation function within the specified time frame

Interaction among management functions within the FMT class is described as well, especially any dependencies. The rationale for not selecting management requirements is explained in Section 8.2 of the ST.

Management requirements may be specified in a PP, especially if the customer has specific operational requirements; however, given the implementation dependence of an ST, it may be preferable to defer the specification of management requirements until the ST is developed.

4.7.2 Security Assurance Measures

This subsection of an ST describes the specific security assurance measures employed to satisfy each SAR specified in the PP. The required EAL is recounted, including any augmentations and extensions. Then, security assurance measures that correspond to each assurance class and family comprising the requisite EAL are discussed in detail. Textual material is supplemented with tables and figures whenever possible to enhance understandability. In particular, a matrix should be prepared that maps SARs to specific security assurance measures. Again, each SAR must map to at least one security assurance measure, while each security assurance measure must map to at least one SAR.

The developer is responsible for explaining how the developer action elements and the content and presentation evidence elements have been achieved for each SAR. The developer is not responsible for addressing evaluator action elements. However, common sense dictates that the developer must ensure that all criteria defined in the evaluator action elements have been satisfied, which is the premise that a Common Criteria Testing Laboratory (CCTL) will use to determine whether or not the ST passes certification. For EAL 2 and below, reference can be made to existing project documentation, such as configuration management procedures, instead of repeating that information in the ST. The caveat, of course, is that the documentation: (1) much exist at the time the ST is submitted for formal evaluation, and (2) project staff actually know and follow the documented procedures.

Exhibit 19 illustrates how the text description of security assurance measures might be written for ADV_FSP.1 in two scenarios: EAL 2 and EAL 3. In the first example, reference is made to existing project documentation to satisfy the requirement. In the second example, the developer action elements and the content and presentation evidence elements are reflected. "Shalls" are replaced by "wills" and other minor editorial changes are made, as necessary, to impart the vendor's security assurance solution.

Exhibit 20 presents a sample matrix that maps SARs to security assurance measures. Because this example is for EAL 2, reference is made to existing project documentation. For EAL 3 and above, reference would be made to specific security assurance measures. This table generally concludes Section 6 of an ST.

Exhibit 19. Sample TSS Security Assurance Measures

Example 1: EAL 2 (reference to existing project documentation)

6.2.3 ADV_FSP.1 — Informal Functional Specification

The ABC functional specification, listed below, implements the security assurance requirements of ADV_FSP.1:

- ABC Functional Specification, version 3.0, dated March 5, 2002

Example 2: EAL 3

6.2.3 ADV_FSP.1 — Informal Functional Specification

6.2.3.1 ADV_FSP.1.1D: As the developer, we will provide a functional specification that is an accurate and complete instantiation of the TSF.

6.2.3.2 ADV_FSP.1.1.C: This functional specification will describe the TSF and all external interfaces.

6.2.3.3 ADV_FSP.1.2C: This functional specification will be internally consistent.

6.2.3.4 ADV_FSP.1.3C: This functional specification will describe the purpose and use of all external TSF interfaces and include a description of exception handling and error messages.

6.2.3.5 ADV_FSP.1.4C: The functional specification will represent the complete TSF.

4.8 Section 7: PP Claims

The seventh section, PP Claims, indicates and explains the degree of conformity between an ST and the referenced PP, which may range from none to complete conformity. All PP Claims must be substantiated by sufficient explanation, justification, and evidence:[19]

> The fundamental requirement is that the ST content be complete, clear, and unambiguous such that evaluation of the ST is possible, the ST is an acceptable basis for the TOE evaluation, and the traceability to any claimed PP is clear.

The placement of PP Claims as the seventh section in an ST is somewhat puzzling. The reader of an ST, whether customer or evaluator, needs to be cognizant of the PP Claims prior to reading Sections 2 to 5 of an ST in order to understand and interpret them correctly. Hence, it would be more logical to have PP Claims as Section 2, rather than Section 7, of an ST. This section is composed of three subsections: PP Reference, PP Tailoring, and PP Additions.

4.8.1 PP Reference

The PP Reference subsection cites the specific PP to which conformance is claimed by stating the full title, version number, and publication date of the PP. Five possible conformance scenarios may be claimed:[19]

Exhibit 20. TSS Security Assurance Mapping

Table x—Mapping Between SARs and Security Assurance Measures (EAL 2)

SAR	Security Assurance Measure
ACM_CAP.2	• Configuration Management and Delivery Documentation • Hardware Functional Specification
ADO_DEL.1	• Configuration Management and Delivery Documentation
ADO_IGS.1	• Installation and Configuration Documentation
ADV_FSP.1	• Configuration Guide • Administrative Guide • System Architecture Documentation • Detailed Design Documentation • Error Messages and Exception Handling Documentation • Quick Installation Guide • Hardware Functional Specification
ADV_HLD.1	• System Architecture Documentation • Detailed Design Documentation • Error Messages and Exception Handling Documentation
ADV_RCR.1	• Correspondence White Paper
AGD_ADM.1	• Configuration Management and Delivery Documentation • Configuration Guide • Administrative Guide • Release Notes • Quick Installation Guide
ATE_COV.1	• Test Plan • Test Procedures • Test Analysis Report
ATE_FUN.1	• Test Plan • Test Procedures • Test Analysis Report
ATE_IND.2	• Test Plan • Test Procedures • Test Analysis Report
AVA_SOF.1	• Administrative Guide
AVA_VLA.1	• Vulnerability Assessment

1. *None* — No claim of conformance is made. Sections 3 to 5 of the ST must be provided in their entirety. Subsections 7.2 and 7.3 of PP Claims are omitted.
2. *Complete* — Complete conformance with the cited PP is claimed. Sections 3 to 5 of the PP need only be referenced, not restated. Subsections 7.2 and 7.3 of PP Claims are omitted.
3. *Complete with tailoring* — Complete conformance with the cited PP is claimed; however, some tailoring of assumptions, security objectives, or security requirements has taken place. Sections 3 to 5 of the PP should be restated and the tailoring highlighted. Subsection 7.2 of PP Claims is included to explain and justify the tailoring.

4. *Complete with additions* — Complete conformance with the cited PP is claimed. However, some assumptions, security objectives, or security requirements have been added. Sections 3 to 5 of the PP should be restated and the additions highlighted. Subsection 7.3 of PP Claims is included to explain and justify the additions.
5. *Partial*— Partial conformance with the cited PP is claimed. This scenario is only valid if the PP is for a composite TOE and the ST reflects one of the component TOEs. It is not valid to claim partial conformance to a PP for a monolithic or component TOE; Security Targets are not written for functional packages because they cannot be subjected to a formal CC/CEM evaluation. The parts of Sections 3 to 5 of the PP that apply to the component TOE are restated. Any tailoring or additions should be highlighted, and Subsections 7.2 and 7.3 of PP Claims should be included to explain and justify the tailoring or additions.

The "none" and "complete" scenarios are mutually exclusive. The "complete with tailoring" and "complete with additions" scenarios are not mutually exclusive. The "partial" scenario may or may not include PP tailoring and PP additions.

4.8.2 PP Tailoring

This subsection of an ST identifies any assumptions from Section 3 of the PP, security objectives from Section 4 of the PP, or security requirements from Section 5 of the PP that were tailored or customized. The valid tailoring options for security requirements are limited to performing element operations that were not performed in the PP, such as assignment, selection, refinement, and iteration. An explanation of what tailoring was performed and why it was performed is provided.

4.8.3 PP Additions

This subsection of an ST identifies any security assumptions, objectives, or requirements that are included in the ST but not the referenced PP. An explanation of what additions were made and why is provided. Exhibit 21 presents a sample PP Claims section.

4.9 Section 8: Rationale

The eighth section, Rationale, provides evidence that the ST is complete, correct, consistent, and coherent, internally to itself and with the corresponding PP. As the standard states:[19]

> This part of the ST presents the evidence used in the ST evaluation. This evidence supports the claims that the ST is a complete and

Exhibit 21. Sample PP Claims

7 PP Claims

This section presents the PP conformance claims.

7.1 PP Reference

This security target conforms to the following PP:

- Application-Level Firewall Protection Profile for Low-Risk Environments, U.S. DoD, Version 1.d.1 (draft) September, 1999.

Degree of conformance claimed:

- Complete with tailoring
- Complete with additions

7.2 PP Tailoring

The SFRs, and SAR listed in the table below were tailored from those stated in the referenced PP.

Tailored Item	Tailoring Conducted	Justification
FAU_GEN.1	refinement	need to identify (1) audit requirement as being minimal, basic, detailed, or not specified and (2) items to be included in the audit.
FAU_SAR.3+1 FAU_SAR.3+2	refinement and iteration	need to clarify that the TOE should be capable of: (1) searching the audit data for user identity, presumed subject address, ranges of dates, ranges of time, and ranges of IP addresses; and (2) sorting audit data based on chronological order or occurrence.
AVA_VLA.1	refinement	need to specify the minimum identified vulnerabilities for which the evaluated TOE must be analyzed.

7.3 PP Additions

The SFR listed in the table below was added to those stated in the referenced PP.

Item Added	Justification
FIA_UAU.5	The referenced PP specifies some of the characteristics required of the two authentication mechanisms (reusable and single-use) that may be used via the FIA_UAU.1 and FIA_UAU.4 SFRs. However, it fails to provide an SFR identifying what types of authentication mechanisms may be used or the conditions requiring their use.

cohesive set of requirements, that a conformant TOE would provide an effective set of IT security countermeasures within the security environment, and that the TOE summary specification addresses the requirements.

In summary, the Rationale demonstrates and proves that:

- Section 4 Security Objectives are responsive to the Section 3 TOE Security Environment.
- Section 5 Requirements implement all Section 4 Security Objectives.
- Section 6 design solution implements all Section 5 Requirements.
- PP compliance and SOF claims are valid.

The Rationale contains four subsections: Security Objectives Rationale, Security Requirements Rationale, TOE Summary Specification Rationale, and PP Claims Rationale.

4.9.1 Security Objectives Rationale

The Security Objectives Rationale subsection demonstrates that the security objectives stated in Section 4 for the TOE and the TOE environment are completely responsive to the TOE security environment described in Section 3 of the ST. In other words, this subsection proves that the security objectives uphold all assumptions, counter all threats, and enforce all organizational security policies. Differences between the PP and ST security objectives rationales should be highlighted, such as:[19]

- Additional or tailored assumptions identified in the ST that are addressed by the Security Objectives
- Additional or tailored threats identified in the ST that are addressed by the Security Objectives
- Tailored organizational security policies identified in the ST that are addressed by the Security Objectives
- Additional or tailored security objectives that counter threats or uphold organizational security policies

Exhibit 22 presents sample security objectives rationales. Example 1 contains the textual discussion of why each TOE security objective is *necessary* and *sufficient* to counter the corresponding threats. An objective may map to one or more threats; likewise, more than one objective may map to the same threat. This discussion is followed by a summary table (Example 2), which depicts the mapping to ensure that all assumptions, threats, and organizational security policies are accounted for. Example 3 presents the textual discussion of why each security objective for the TOE environment is *necessary* and *sufficient* to counter the corresponding threats. In this example, two objectives map to the same threat and several assumptions are restated to give them the force of security objectives. Again, the textual discussion is followed by a summary table (Example 4).

The Security Objectives Rationale must also explain and justify why any assumptions, threats, organizational security policies, and security objectives contained in the PP are either tailored or not included in the ST. Valid technical or operational reasons for the modifications or exclusions must be provided. Any new assumptions, threats, or security objectives must be justified as well. Example 5 illustrates two possible scenarios.

4.9.2 Security Requirements Rationale

The purpose of the Security Requirements Rationale subsection is to prove that the combination of the security requirements for the TOE and those for

Exhibit 22. Security Objectives Rationale

Example 1: Rationale for Target of Evaluation (TOE) Security Objectives

- O1 This security objective is necessary to counter threat T1 (replay). O1 requires the TOE to prevent the reuse of authentication data — even if valid authentication data is obtained, it cannot be used to mount an attack.
- O2 This security objective is necessary to counter threats T2 and T3 (spoofing and residual information). O2 requires that: (a) all TOE information flows be mediated *a* priori, and (b) no residual information be transmitted internal or external to the TOE.
- O3 This security objective is necessary to counter theat T4 (TSF attack). O3 requires the TOE to protect itself from attempts to bypass, deactivate, or tamper with TOE security functions.

Example 2: Mapping Security Objectives to Threats

Security Objective	Threat			
	T1	T2	T3	T4
O1	X			
O2		X	X	
O3				X

Example 3: Rationale for TOE Environment

- O4 This non-IT security objective is necessary to counter threat T5 (usage). O4 requires that the TOE be delivered, installed, administered, and operated in a secure manner.
- O5 This non-IT security objective is necessary to counter threat T5 (usage). O5 requires that authorized administrators and users receive appropriate training on a regular basis about: (a) the secure use of the TOE, and (b) any residual risk.

The remaining security objectives are a restatement of the assumptions for the TOE security environment.

Example 4: Mapping security objectives for the environment to threats and assumptions

Security Objective	Threat	Assumptions		
	T5	A17	A18	A21
O4	X			
O5	X			
O6		X		
O7			X	
O8				X

Example 5: Rationale for Exceptions

- 8.1.x Three assumptions for the TOE security environment (A15, A19, A20) were modified in this ST. The refined assumptions were necessary because of the specific TOE hardware and software platforms. However, the modified assumptions maintain the original intent of the PP.
- 8.1.x Security objective O23 and threat T17 from the PP are not applicable to this ST because the only human users of the TOE are the authorized system administrators. There are no end-users.

the TOE environment satisfy and correspond to all security objectives. In particular, the following must be demonstrated:[19,22]

- The composite security requirements are necessary, sufficient, and mutually supportive.
- The collection of individual security functional and assurance components for the TOE and the TOE environment (IT and non-IT) meet all stated security objectives.
- The selection of specific SFRs and SARs is justified, in particular: (a) the use of explicit requirements, (b) EAL augmentations and extensions, and (c) non-satisfaction of dependencies.
- The specified and claimed SOF is consistent with TOE security objectives.

The Security Requirements Rationale is generally organized in three parts that prove that individually and collectively specified security requirements are necessary, sufficient, and mutually supportive. Differences between PP and ST security objectives or requirements should be noted during the generation of these proofs, such as new or tailored objectives or requirements. Each of these three parts of the proof is discussed below.

To prove the necessary criteria, the developer must demonstrate that each requirement (SFR and SAR) is necessary and that no redundant or extra requirements are included; all requirements must correspond to a security objective. An easy way to do this is through a table that cross-references objectives to requirements.

Next, the sufficiency criteria are proven. Formal arguments are constructed to explain why security requirements are sufficient to satisfy security objectives. All security requirements are examined: SFRs, SARs, standard, explicit, principal, supporting, IT, IT environment, and non-IT environment. Particular attention is paid to element operations and iterations: how and why they were performed. Part of the sufficiency analysis verifies that the specified level for events that can be audited has been achieved (see Exhibits 23 and 24). The sufficiency criteria for SARs are assessed to determine whether or not the EAL is adequate and any augmentations or extensions are appropriate. The sufficiency argument must prove that the specified EAL is (1) neither too strong nor too weak, and (2) technically feasible, given the implementation specifics of the TOE and the current state of technology. Exhibit 25 illustrates a rationale for EAL 2.

The third part of Security Requirements Rationale is the proof that the security requirements are mutually supportive. As the standard states:[22]

> IT security requirements are complete and internally consistent by demonstrating that they are mutually supportive and provide an integrated and effective whole.

The mutually supportive proof is constructed through a three-step analytical process:

Exhibit 23. Requirements Rationale — SFRs Necessary

8.2 Security Requirements Rationale

8.2.1 SFRs Necessary

This ST satisfies the SFRs specified in the PP as demonstrated in the text and table below.

FDP_IFC.1+1. Subset information flow control

This component identifies the entities involved in the unauthenticated information flow control TSP that are using all network services except HTTP, SMTP, FTP, and Telnet. This component traces back to and aids in meeting security objective O21.

FDP_IFC.1+2. Subset information flow control

This component identifies the entities involved in the authenticated information flow control TSP. This component traces back to and aids in meeting security objective O21.

FDP_IFF.1+1. Simple security attributes

This component identifies the attributes of users sending and receiving information in the unauthenticated TSP, as well as the attributes for the information itself. The policy is enforced by permitting or restricting information flow. This component traces back to and aids in meeting security objective O22.

FDP_IFF.1+2. Simple security attributes

This component identifies the attributes of users sending and receiving information in the authenticated TSP, as well as the attributes for the information itself. The policy is enforced by permitting or restricting information flow. This component traces back to and aids in meeting security objective O.22.

Table 1 summarizes the mapping between SFRs and security objectives.

Table 1 SFRs Mapped to Security Objectives

SFR	Security Objective O21	Security Objective O22
FDP_IFC.1+1	X	
FDP_IFC.1+2	X	
FDP_IFF.1+1		X
FDP+IFF.1+2		X

1. The resolution or non-resolution of component dependencies is analyzed.
2. The internal consistency between security requirements is analyzed.
3. The proactive resistance of SFRs to attacks is analyzed.

The component dependency analysis examines all three possible combinations of dependencies: SFR to SFR, SFR to SAR, and SAR to SAR. The resolution of each dependency for each SFR and all iterations of that SFR are investigated. The developer must prove which dependencies were satisfied and identify those that were not. A justification must be provided for all unsatisfied dependencies, which explains why it was not necessary to incorporate the supporting requirement. Potential reasons might be that the implementation specifics contained in the ST made the requirement unnecessary,

Exhibit 24. Requirements Rationale: Auditable Events

8.2 Security Requirements Rationale

8.2.x Auditable Events Rationale

The auditable events provided by ABC were reviewed against the auditable events for the minimal or basic level of audit for the functional requirements. It was found that ABC provided auditable events for the applicable functionality in all areas except for export, import, confidentiality, and integrity. It was decided that it would not be appropriate for ABC to audit these activities since all user data messages sent between two ABCs: (a) have an integrity check applied, (b) are encrypted for confidentiality, and (c) are imported and exported from both ABCs. These are routine events for ABC and hence not appropriate for auditing. Therefore, "not specified" was selected for the level of audit and all pertinent auditable events were listed.

or that the function of the supporting requirement was satisfied by a requirement for the IT or non-IT environment. Results of the component dependency analysis are presented in text supplemented by a tabular summary, as shown in Exhibit 26.

The internal consistency analysis demonstrates that there are no overlapping, conflicting, or ambiguous requirements. Audit requirements, management requirements, component iteration, and element operations are all within the scope of the internal consistency analysis. If the ST is for a component TOE, STs for related component TOEs should also be evaluated as part of the internal consistency analysis. Results of this analysis are presented in text supplemented by a tabular summary.

The analysis of the proactive resistance of SFRs to attacks focuses on four key parameters:[22]

- Bypassability
- Tampering
- Deactivation
- Detection

The standard defines an SFR to counter bypassability:[20]

Exhibit 25. Requirements Rationale: SARs necessary and sufficient

8.2 Security Requirements Rationale

8.2.2 SARs Necessary and Sufficient

This ST satisfies the SARS specified in the PP. Section 5.2 of this document identifies the Configuration Management, System Delivery Procedures, System Development Procedures, Guidance Documents, Testing, and Vulnerability Analysis measures applied to satisfy the EAL 2 requirement.

EAL 2 was chosen to provide a low to moderate level of independently assured security. The chosen assurance level is consistent with the postulated threat environment — the threat of malicious attacks is moderate and the product has undergone a search for obvious flaws.

FPT_RVM.1 Nonbypassability of the TSP

FPT_RVM.1.1	*The TSF shall ensure that TSP enforcement functions are invoked and succeed before each function within the TSC is allowed to proceed.*

For this parameter, the analysis of proactive resistance is straightforward:[22]

- Is FPT_RVM.1.1 included in the security requirements?
- Does FPT_RVM.1.1 operate in an always invoked status?
- Within the context of the TOE's implementation dependence and operational environment, are any other SFRs necessary to ensure that FPT_RVM.1.1 is successful?

The standard defines several SFRs to counter physical and cyber tampering:[20]

- The FPT_SEP family defines requirements for logical domain separation to prevent "external interference or tampering by untrusted subjects."
- The FTP_PHP family defines requirements for physical protection of the TSF to "detect/resist physical tampering."
- The FMT class defines requirements to "(a) restrict the ability to modify security attributes or configuration data, and (b) protect the integrity of security integrity data."
- The FTP_TRP family defines requirements for trusted paths to resist spoofing attacks.

The analysis of proactive resistance is straightforward for this parameter as well:

- Have the appropriate SFRs been specified to prevent physical and cyber tampering?
- Do these SFRs operate in an always invoked status?

The next step in analyzing the proactive resistance of SFRs is to determine which SFRs are likely to be subject to deactivation attacks. Two examples from the standard include:[20]

FAU_STG.2 Guarantees of audit data availability

FAU_STG.2.1	*The TSF shall protect the stored audit records from **unauthorized deletion.***
FAU_STG.2.2	*The TSF shall be able to [selection: prevent, **detect]** modifications to audit records.*
FAU_STG.2.3	*The TSF shall ensure that [assignment: metric for saving audit records] audit records will be maintained when the following conditions occur: [selection: audit storage exhaustion, failure, **attack].***

Exhibit 26. Requirements Rationale: Component Dependency Analysis

8.2 Security Requirements Rationale

8.2.2 SARs Necessary and Sufficient

Table 1 depicts dependencies between SFRs. For completeness, all SFRs are listed in the table whether or not they have a dependency. As shown, all dependencies are satisfied with the exception of FMT_MSA.3. This functionality is provided by two security requirements for the IT environment: ITENV.1 Setting User Attributes and ITENV.2 Modifying TSF Data. In two cases a dependency is satisfied by a component that is hierarchical to the required component: FDP_IFF.1 is satisfied by FDP_IFF.2, and FIA_UID.1 is satisfied by FIA_UID.2. This is indicated in the table by an "H" after the reference line number.

Table 1. Resolution of SFR-to-SFR Dependencies

Ref.	SFR	Name	Dependencies	Resolution
1	FAU_GEN.1	Audit data generation	FPT_STM.1	22
2	FAU_SEL.1	Selective audit	FAU_GEN.1 FMT_MTD.1	1 14
3	FDP_ACC.1	Subset access control	FDP_ACF.1	4
4	FDP_ACF.1	Security attribute based access control	FDP_ACC.1 FMT_MSA.3	3 ITENV.1 ITENV.2
5	FDP_ETC.1	Export of user data without security attributes	FDP_ACC.1 or FDP_IFC.1	3 —
6	FDP_IFC.1	Subset information flow control	FDP_IFF.1	7H
7	FDP_IFF.2	Hierarchical security attributes	FDP_IFC.1 FMT_MSA.3	— ITENV.1 ITENV.2
8	FDP_ITC.1	Import of user data without security attributes	FDP_ACC.1 or FDP.IFC.1 FMT_MSA.3	— 6 ITENV.1 ITENV.2
9	FDP_UCT.1	Basic data exchange confidentiality	FTP_ITC.1 or FTP_TRP.1 FDP_ACC.1 or FDP_IFC.1	24 — 3 —
10	FDP_UIT.1	Data exchange integrity	FDP_ACC.1 or FDP_IFC.1 FTP_ITC.1	3 — 24
11	FIA_ATD.1	User attribute definition	None	—
12	FIA_UAU.2	User authentication before any action	None	—
13	FIA_UID.2	User identification before any action	None	—
14	FMT_MTD.1	Management of TSF data	FMT_SMR.1	17
15	FMT_REV.1	Revocation	FMT_SMR.1	17
16	FMT_SAE.1	Time-limited authorization	FMT_SMR.1 FPT_STM.1	17 22
17	FMT_SMR.1	Security roles	FIA_UID.1	13H
18	FPT_AMT.1	Abstract machine test	none	—
19	FPT_ITI.1	Inter-TSF detection of modification	none	—
20	FPT_RVM.1	Non-bypassability of the TSP	None	—
21	FPT_SEP.1	TSF domain separation	None	—
22	FPT_STM.1	Reliable time stamps	None	—
23	FPT_TDC.1	Inter-TSF basic TSF data consistency	None	—
24	FTP_ITC.1	Inter-TSF trusted channel	None	—

FMT_MOF.1 Management of security functions behavior

FMT_MOF.1.1	*The TSF shall restrict the ability to [selection: determine the behavior of, **disable**, enable, **modify** the behavior of] the functions [assignment: list of functions] to [assignment: the authorized identified roles].*

Once these SFRs have been identified, an assessment is made as to whether or not the protection afforded them is sufficient to counter the anticipated deactivation attacks.

The last step to analyzing the proactive resistance of SFRs is to determine: (1) the extent of detection needed to detect attacks and accidental or intentional misconfiguration of the TSF, and (2) the sufficiency of the proposed detection functions.[22] The sufficiency criteria may be satisfied by an SFR alone or a collection of SFRs.Two examples from the standard include:[20]

For all four parameters, the results of the proactive resistance analysis are presented in text, supplemented by diagrams and a tabular summary.

FDP_SDI.2	*Stored data integrity and action*
FDP_SDI.2.1	*The TSF shall monitor user data stored within the TSC for [assignment: integrity errors] on all objects, based on the following attributes [assignment: user data attributes].*
FDP_SDI2.2	*Upon **detection** of a data integrity error, the TSF shall [assignment: action to be taken].*

4.9.3 TOE Summary Specification Rationale

The third major part of the ST Rationale is the TSS Rationale. As the standard states, the purpose of this subsection is to:[19]

> ...show that the TOE security functions and assurance measures are suitable to meet the TOE security requirements.

In particular, three items must be demonstrated:[19]

1. The proposed security solution satisfies all SFRs in the ST.
2. SOF claims are valid.
3. Proposed security assurance measures satisfy all SARs in the ST.

All SFRs are within the scope of the proof, those carried forward from the PP and those added or tailored by the ST; any SFRs deleted by the ST are not included. Likewise, all SARs are within the scope of the proof, including augmentations and extensions. The same criteria are used to generate the TSS Rationale as the Security Requirements Rationale: necessary, sufficient, and mutually supportive.

Subsection 8.1 of the Rationale proves the consistency, correctness, completeness, and coherence between Security Objectives and the TOE Security

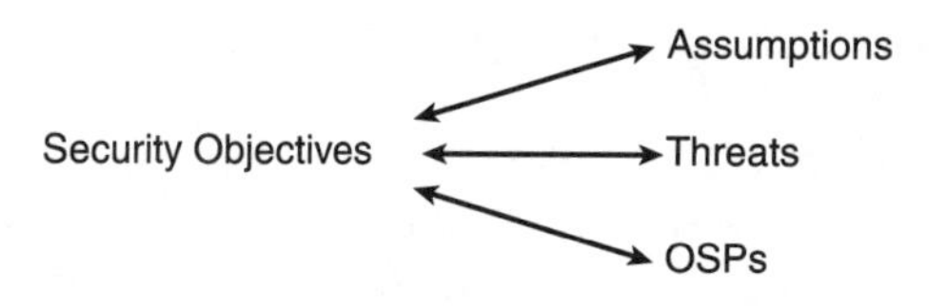

Exhibit 27. Subsection 8.1 of the Rationale

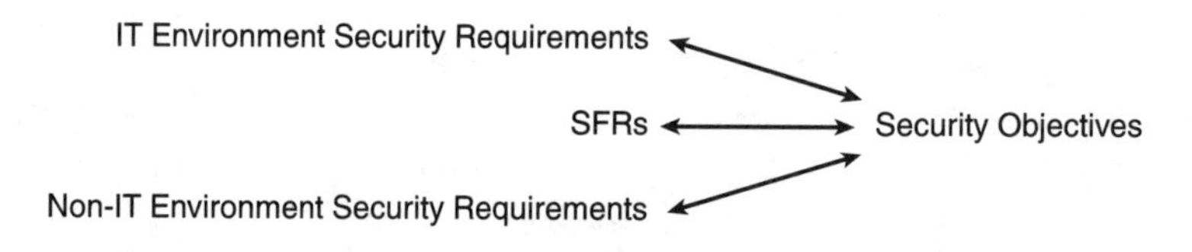

Exhibit 28. Subsection 8.2 of the Rationale

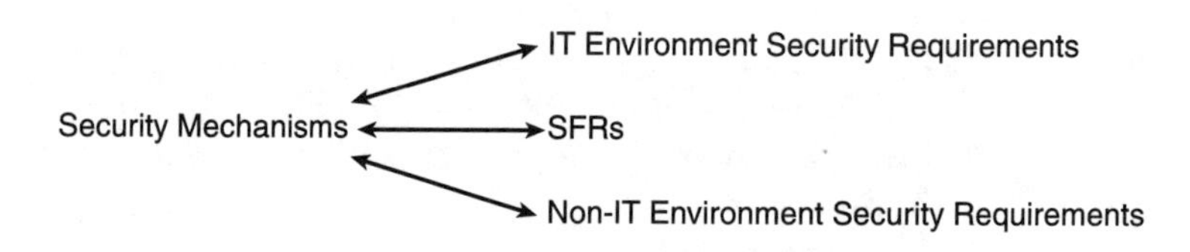

Exhibit 29. Subsection 8.3 of the Rationale

Environment Assumptions, Threats, and Organizational Security Policies (see Exhibit 27). Subsection 8.2 of the Rationale proves the consistency, correctness, completeness, and coherence (1) between Security Objectives and Security Requirements, and (2) among Security Requirements (see Exhibit 28). Subsection 8.3 of the Rationale completes the picture by providing the missing link: the consistency, correctness, completeness, and coherence (1) between Security Mechanisms and Security Requirements, and (2) among Security Mechanisms (see Exhibit 29).

The same process is used to demonstrate correspondence between security mechanisms and security requirements that was used to demonstrate correspondence between security objectives and security requirements. Each SFR is mapped to the specific security mechanism that implements the specified functionality. All SFRs must map to at least one mechanism, while each security mechanism must map to at least one SFR. There should not be any unmapped SFRs or security mechanisms. The mapping is performed to ensure that: (1) no new unspecified functionality could create a vulnerability, (2) no SFRs are missing from the implementation, (3) no vulnerabilities were introduced by decomposing a requirement into the level of detail necessary for implementation, and (4) the proposed solution is sufficiently robust and resilient.

Exhibit 30. Requirements Rationale: TOE SOF Claims

8.3. TSS Rationale
8.3.x. SOF Claims
The PP required an overall SOF of SOF-basic. This ST claims: (a) the SOF requirement is valid, and (b) conformance with this SOF requirement. The identified metrics and SOF claim is commensurate with EAL 2. If the rules specified in the TSS governing passwords are followed, the probability of guessing a password is less than one in a million. Also, the random number generator used to develop password sets complies with the "Statistical random number generator tests" and "Continuous random number tests" found in Section 4.11.1 of FIPS PUB 140-2. This ensures that the requirements of the AVA_SOF.1 assurance requirement are met by the implementation defined in this ST.

The necessary criteria are demonstrated by representing the mapping in tabular form. The sufficient and mutually supportive criteria are explained in text. Again, the goal is prove that the composite security mechanisms represent an integrated and effective whole, using the Security Requirements Rationale as input.[19,22]

As a corollary to the sufficiency analysis, a rationale is also generated for SOF claims. This argument proves that the minimum and any explicit SOF claims are consistent with the security objectives and resistant to all foreseeable types of attacks. Exhibit 30 illustrates a rationale for SOF–basic.

A similar procedure is followed to demonstrate that security assurance measures are necessary, sufficient, and mutually supportive. In this case, security assurance measures are mapped to SARs (see Exhibit 31).

This mapping is summarized in tabular form to demonstrate the necessary criteria. Text is necessary to explain the sufficient and mutually supportive criteria and, hence, justify that the proposed security assurance measures will achieve (and sustain, if appropriate) the specified EAL.

4.9.4 PP Claims Rationale

The fourth major part of the Rationale is the PP Claims Rationale, which proves that the PP claims made in the TSS are valid. Compliance between the ST and the referenced PP is demonstrated by proving that:

- All PP Security Objectives are included in the ST.
- All refinements or additions to PP Security Objectives are valid.
- All PP Security Requirements are included in the ST.
- All refinements, additions, and element operations performed are valid.

Security Assurance Measure ←——→ SAR

Exhibit 31. Security Assurance Measures Mapped to SARs

If the TSS PP claim is "none," this part of the Rationale is marked not applicable. If the TSS PP claim is "complete," the proof is straightforward and the developer responds to the first and third bullets. If the TSS PP claim is "complete with tailoring," "complete with additions," or "partial," then more detail is required and the developer responds to all four bullets.

4.10 Summary

A Security Target (ST), a combination of security objectives, functional and assurance requirements, summary specifications, PP claims, and rationales, is an *implementation-dependent* response to a PP that is used as the basis for development and evaluation of a TOE.[19,24,112] In other words, the PP specifies security functional and assurance requirements, while an ST provides a detailed design that incorporates specific functional security mechanisms and security assurance measures to fulfill these requirements.

Several stakeholders interact with an ST. STs are written by developers in response to a PP, read by potential customers, and reviewed and assessed by evaluators. After approval, STs are used by developers as the foundation for constructing a TOE. Once written, an ST is not cast in concrete; rather, it is a living document. The CC/CEM and artifacts map to generic system lifecycle and procurement phases. An ST corresponds to the design phase — a design is generated by a developer in response to security requirements stated by a customer in a PP and the quality of this design is verified through the ASE class security assurance activities. An ST is part of pre-award procurement activities; STs are included in proposals submitted by potential offerors and evaluated by the source selection team.

An ST is a formal document with specific content, format, and syntax requirements. This formality is imposed to ensure that STs are accurately and uniformly interpreted by all the different stakeholders. An ST is not written *per se*; rather, it captures the culmination of a series of analyses conducted by the developer to create a solution to a customer's security requirements. An ST is a cohesive whole; as such, interaction among the eight required sections is extensive.

Section 1 introduces an ST by identifying its nature, scope, and status. Section 2 describes the system type, architecture, and logical and physical security boundaries. The System Type subsection introduces the TOE by defining the type of technology it represents. A main objective of the Architecture subsection is to differentiate the TOE and TOE security architectures. The TOE architecture defines the context in which the TOE security function must operate. The TOE security architecture defines an implementation specific instance of the TSF within the constraints imposed by the TOE architecture and the asset types and sensitivities defined in the PP. If the ST is for a security-specific COTS product, the TOE and security architectures may be the same. In contrast, if the ST is for a component TOE that is part of a system, major distinctions between the TOE and security architectures may exist. The physical security boundary represents the perimeter within which TOE security functions are effective. IT entities within this perimeter are under the TSF scope of control and are protected to the extent and in the manner specified. IT entities outside this are not within the TSC and may or may not be trusted. The logical security boundary is defined in terms of security services or features.

Section 3 states the assumptions, analyzes the threats, and cites organizational security policies that are applicable to the TSF. The ST Security Environment section mirrors the PP Security Environment section to the extent appropriate; differences between the two result from the fact that a PP is implementation independent while an ST is implementation dependent. The developer has four options concerning assumptions:

1. PP assumptions may simply be restated verbatim.
2. PP assumptions may be modified or tailored to reflect TOE implementation specifics.
3. New assumptions may be added.
4. PP assumptions may not be carried forward if they are inappropriate or not applicable.

The developer determines which threats contained in the PP are:

- Addressed by the TOE design solution defined in the ST
- Addressed by the TOE security environment defined in the ST
- Not applicable to either the TOE or TOE environment

Likewise, the developer determines the residual risk of each threat after the countermeasures contained in the design solution (technical, operational, and procedural) have been deployed. OSPs are unique to each organization and its mission and assets; the customer is the owner of OSPs. As a result, the developer is limited in terms of what actions can be performed with PP OSPs.

Section 4 delineates security objectives for the TOE and the IT environment. These objectives are derived from an analysis of the assumptions, threats, and security policies articulated in Section 3. The developer responds to the security objectives in the applicable PP. A determination is made whether or not: (1) the security objectives in the PP are valid, and (2) any new security objectives must be added.

Section 5 implements security objectives expressed in Section 4 through a combination of SFRs and SARs. These SFRs and SARs are derived from an analysis of the system architecture and security boundaries stated in Section 2 and the perceived risk of compromise presented in Section 3. The developer responds to the requirements in the PP. Requirements in the PP may be simply restated or the developer may add implementation specific details by:

- Completing element operations that were not performed in the PP
- Resolving component dependencies that were not resolved in the PP
- Refining or iterating functional components to reflect the proposed security solution

- Specifying audit requirements not supplied in the PP
- Adding new SFRs that are necessary because of the implementation-dependent nature of the ST
- Reassigning SFRs to security requirements for the TOE or non-IT environment
- Omitting SFRs if they are unnecessary, redundant, or conflicting

Section 6, the TOE Summary Specification (TSS), defines:[19,22]

- Specific IT security functions that satisfy each SFR identified in Section 5
- Exact security mechanisms or techniques that are used to implement each IT security function
- Precise security assurance measures that are employed to satisfy each SAR identified in Section 5

A six step process is followed to elucidate the TSF:

Step 1 — TSF packages are mapped to SFRs.
Step 2 — The TSF structure is defined.
Step 3 — Specific security mechanisms that implement security functions are identified.
Step 4 — The capture of audit requirements is explained.
Step 5 — Satisfaction of SOF requirements is demonstrated.
Step 6 — The implementation of management requirements is described.

Section 7 clarifies which PP the ST was developed in response to and explains the degree of conformity between an ST and the referenced PP, which can be none, complete, complete with tailoring, complete with additions, and partial.

The last section, Section 8, proves that:

- Security objectives stated in Section 4 uphold all assumptions, counter all threats, and enforce all security policies identified in Section 3.
- Requirements specified in Section 5 implement all security objectives stated in Section 4.
- The design solution presented in Section 6 implements all requirements stated in Section 5.
- Claims made about PP compliance and achievement of the specified strength of function are valid.

The Rationale proves that the ST is complete, correct, consistent, and coherent, internally to itself and with the referenced PP. The developer demonstrates that the security objectives, security requirements, and TSS are necessary, sufficient, and mutually supportive.

4.11 Discussion Problems

1. What operations can and cannot be performed in an ST?
2. Discuss the pros and cons of deferring the resolution of dependencies until development of an ST.
3. Explain the relationship between Security Targets and monolithic, component, and composite TOEs.
4. How do TOE boundaries in a PP relate to logical and physical security boundaries in an ST?
5. How do customers and evaluators use the assumptions stated in Section 2.1 of an ST?
6. How are threats stated in Section 2.2 of an ST characterized? How is this different from the characterization of threats in a PP?
7. What makes an ST unique compared to other detailed design specifications?
8. Where is SOF mentioned in an ST, and why?
9. Explain the relationship among the TOE, TSF, TSC and TSFI for (a) component TOE, (b) composite TOE, (c) COTS product, and (d) system.
10. Develop a rationale for the SOFs described in Examples 1 and 3 of Exhibit 18.
11. Describe the similarities and differences between the TOE and security architectures.
12. What is the purpose of the security boundary in an ST?
13. What is the purpose of management requirements?
14. What criteria are used to prove that security requirements are: (a) necessary, (b) sufficient, and (c) mutually supportive?
15. What is the difference between the TSS and the TSS Rationale?
16. How do you prove that an SOF requirement is valid?
17. How do you prove that an SOF requirement has been met?
18. Can a PP claim be both complete with tailoring and complete with additions? Why?
19. What does the component dependency analysis prove? What components are within the scope of the component dependency analysis?
20. What is the relationship between (a) SFRs and FPs, (b) SFRs and security mechanisms, (c) SFRs and security assurance measures, and (d) SARs and EALs?

Chapter 5

Verifying a Security Solution: Security Assurance Activities

This chapter explains how to verify a security solution, whether a system or commercial "off the shelf" (COTS) product, using the Common Criteria/Common Evaluation Methodology (CC/CEM). The conduct of security assurance activities is examined in detail, in particular why, how, when, and by whom these activities are conducted. Guidance is provided on how to interpret the results of security assurance activities. The relationship between these activities and a generic system lifecycle, a generic procurement process, and system certification and accreditation (C&A) is explained. Finally, the roles of security assurance activities and ongoing system operations and maintenance are highlighted.

5.0 Purpose

Security assurance provides confidence that a product or system will meet, has met, or is continuing to meet its stated security objectives.[19] Evidence is generated at major milestones to indicate whether or not a project is on track for achieving and sustaining security objectives. As shown in Exhibit 1, security assurance activities are ongoing throughout the system development lifecycle, from the initiation of a Protection Profile (PP) to the certification of a target of evaluation (TOE); verification is not a one-time event that occurs after a system or product is developed. Security assurance activities also continue during operations and maintenance to ensure that the evaluation assurance level (EAL) to which a system or product has been certified is maintained after initial certification and between recertification cycles.

Exhibit 1. Mapping of CC/CEM Artifacts to Generic System Lifecycle and Procurement Phases

CC/CEM Artifacts and Activities	Generic System Lifecycle Phases	Generic Procurement Phases
none	Concept	• Concept definition • Feasibility studies, needs analysis • Independent cost estimate
• Protection Profile (PP)	Requirements analysis and specification	Request for proposal (tender) issued by customer
• Security assurance activity: APE		
• Security Target (ST)		
• Security assurance activity: ASE	Design	• Technical and cost proposals submitted by vendors • Technical and cost proposals evaluated by customer
• Target(s) of Evaluation (TOE) developed by winning vendor	Development	Contract award
• Security assurance activities: ACM, ADV		
Security assurance activities: ATE, AVA	Verification	• Acceptance of delivery orders • ECPs issued to correct deficiencies in requirements, design, or development
Security assurance activities: ADO, AGD	Validation, installation and checkout	Deployment
Security assurance activities: ALC, AVA, AMA	Operations and maintenance	Transition to maintenance contract
none	Decommissioning	Contract expiration

The customers are in the driver's seat when it comes to security assurance — they determine to what evaluation assurance levels (EALs) products or systems will be evaluated. They also specify any augmentations or extensions necessary to standard assurance packages. The developer and evaluator have action elements to perform and must adhere to the content and presentation of evidence criteria.

The organization and conduct of security assurance activities is systematic and methodical. The design, development, operation, and maintenance of a TOE are evaluated from multiple facets and against the customer's requirements articulated in a PP. Security assurance requirements (SARs) are invoked to ensure that all security functional requirements (SFRs) for the TOE, the IT environment, and the non-IT environment: (1) have been implemented, (2) have been implemented correctly, and (3) are sufficiently robust and resilient to counter identified threats.

Security assurance activities, action elements, and work units are standardized through ISO/IEC 15408-3, the CEM, and the CC evaluation schemes promulgated by the National Evaluation Authorities. This standardization ensures that evaluation results are consistent, impartial, objective, and repeatable regardless of where or by whom the evaluation is performed. The Common Criteria Recognition Agreement (CCRA), the instrument that enforces this uniformity, defines its charter to be:[23]

> a) to ensure that evaluations of Information Technology (IT) products and Protection Profiles are performed to high and consistent standards, and are seen to contribute significantly to confidence in the security of those products and profiles;
> b) to improve the availability of evaluated, security-enhanced IT products and Protection Profiles;
> c) to eliminate the burden of duplicating evaluations of IT products and Protection Profiles; and
> d) to continuously improve the efficiency and cost effectiveness of the evaluation and certification/validation process for IT products and Protection Profiles.

Consequently, when customers receive CC Certificates from a National Evaluation Authority, they can be confident that:[23]

> ...the evaluation and certification/validation processes have been carried out in a duly professional manner:
>
> a) on the basis of accepted IT security evaluation criteria;
> b) using accepted IT security evaluation methods;
> c) in the context of an evaluation and certification/validation scheme managed by a compliant National Evaluation Authority in the participant's country, and
> d) that the CC Certificates authorized and the certification/validation reports issued satisfy the objectives of [the CCRA].

In the past, stand-alone security assessments were made of products or systems after they were built — attempts were made to "test" security into a system. To be sure, flaws were discovered, but this approach caused an inefficient use of time and resources. Two main problems with this approach were:

1. Security has to be specified, designed, built, and verified into a product or system from day one; it cannot be retrofitted.
2. Security testing and evaluation conducted in the blind are not meaningful and yield subjective or ambiguous results. An objective set of criteria (SFRs, SARs, and an EAL) must be in place, against which the security testing and evaluation are conducted.

In contrast, the CC/CEM promotes an incremental verification strategy (PP → ST → FP → component TOEs → composite TOE) based on objective criteria that yield consistent and repeatable results. This approach is similar to that followed to verify safety-critical embedded software systems (see IEC 61508).

Three sources define what security assurance activities are to be performed, how they are to be performed, and by whom. Each of these sources adds an extra level of detail, building upon the information provided by the predecessor document. The three sources are ISO/IEC 15408-3, CEM parts 1 and 2, and CC evaluation schemes promulgated by each National Evaluation Authority.

5.1 ISO/IEC 15408-3

ISO/IEC 15408-3 defines 10 assurance classes, 42 assurance families, and 93 assurance components. Elements that form assurance components are characterized as developer action elements, content and presentation of evidence criteria, and evaluator action elements. A customer chooses from these SARs when formulating Section 5.2 of a PP. ISO/IEC 15408-3 also defines seven standard assurance packages: EAL 1 through 7. As mentioned in Chapter 3, 28 assurance components are not assigned to an EAL. One family, APE, is used to evaluate PPs and is not part of an EAL. Another family, ASE, is used to evaluate STs and also is not part of an EAL. A third family, AMA, is used to maintain the EAL after the initial certification and between recertification cycles. The remainder of the unassigned components is available for use as extensions when customizing an EAL.

The assurance classes, families, and components defined in ISO/IEC 15408-3 form:[19]

> ...a common basis for evaluation of the security properties of IT products and systems [so that there is] comparability between the results of different independent evaluations ... [and] the results are meaningful to a wider audience.

Likewise, the standardized definitions in ISO/IEC 15408-3 permit the customer, developer, and evaluator to all understand beforehand the exact criteria by which a PP, security target (ST), and TOE will be evaluated.

ISO/IEC 15408-3 cites the three primary sources of security vulnerabilities:[21]

1. Incomplete, incorrect, inconsistent, conflicting, or ambiguous requirements
2. Design and construction that (a) do not comply with the requirements specification, (b) are incorrect implementations or interpretations of requirements, or (c) have followed sloppy lifecycle processes
3. Operational security procedures that are either not followed or inadequate relative to the TOE and its operational environment

The CC methodology was created to prevent or mitigate these vulnerabilities. Correctly specified, SFRs and SARs are designed to reduce the likelihood and severity of the consequences of a security failure or compromise.

A CC evaluation consists of:[21]

> ...an active investigation of an IT product or system that is to be trusted ... which measures the validity of the documentation and resulting IT product or system by expert evaluators with increasing emphasis on scope, depth, and rigor.

Evaluators may use a variety of techniques when conducting a CC evaluation, such as:[21]

- Analysis and checking of processes and procedures
- Checking that processes and procedures are being applied
- Analysis of the correspondence between TOE design representations
- Analysis of the TOE design representation against the requirements
- Verification of proofs
- Analysis of guidance documents
- Analysis of functional tests developed and the results provided
- Independent functional testing
- Analysis for vulnerabilities (including flaw hypothesis)
- Penetration testing

Exhibit 2 aligns security assurance classes and evaluation techniques with the vulnerability source(s) they prevent or mitigate. As expected, the majority of the classes and techniques focus on the correct design and construction of a TOE. In addition, the specification of audit and management requirements helps minimize operational security vulnerabilities.

An important observation should be made concerning APE and ASE. The APE and the ASE classes do not contain a family that evaluates the Rationales. Families are defined that evaluate all sections of a PP or ST (e.g., INT, Introduction; DES, TOE Description; ENV, Security Environment; OBJ, security objectives; SRE, security requirements; etc.), except the Rationales. This seems rather puzzling because the Rationales prove that the requirements and TSS are (1) complete, consistent, correct, coherent, and unambiguous; and (2) accurately implement stated security objectives. However, the reason for this omission is simple: each presentation of evidence elements for the security objectives, security requirements, TOE summary specification (TSS), and PP Claims contains criteria for the applicable subsection of the Rationale. Furthermore, the "real" proof that the Rationales are accurate is generated during evaluation of the TOE.

5.1.1 EALs

The seven predefined EALs represent a continuum on the security assurance scale, from EAL 1 (lowest) to EAL 7 (highest). Augmentations and extensions

Exhibit 2. Mapping between Vulnerability Sources, Security Assurance Classes, and Evaluation Techniques

	Vulnerability Source		
	Requirements	Design and Construction	Operation
Security Assurance Class	APE, ASE	• ASE, ACM, ADO, ADV, AGD, ALC, ATE, AVA	• ACM, AGD, ALC, AVA, AMA • audit requirements • management requirements
Evaluation Techniques	• verification of proofs	• analysis and checking of processes and procedures • checking that processes and procedures are being applied • analysis of the correspondence between TOE design representations • analysis of TOE design representation against requirements • verification of proofs • analysis of guidance documents • analysis of functional tests developed and the results provided • independent testing • analysis for vulnerabilities • penetration testing	• analysis and checking of processes and procedures • checking that processes and procedures are being applied • analysis of guidance documents • analysis for vulnerabilities • penetration testing

may be used to create customized EALs, such as an ~EAL 3.5. EALs only apply to TOEs; they do not apply to PPs or STs.

EAL 1, referred to as functionally tested, provides a minimum level of confidence in the correct operation of the TSF.[21] EAL 1 is appropriate for environments where no serious security threats are anticipated. As shown in Exhibit 3, EAL 1 consists of seven assurance components, all of which are the lowest hierarchy. These seven components establish a basic assurance capability. Ironically, ATE_FUN.1, functional testing, is not included despite this EAL's name.

EAL 2, referred to as structurally tested, provides a low to moderate level of confidence in the correct operation of the TSF.[21] EAL 2 is intended to represent normal commercial practices for the development of non-security-critical products and systems. EAL 2 is often assigned to legacy systems that must be certified long after they were developed and, as a result, little design documentation exists. Stated another way, it is extremely difficult, if not impossible, to certify legacy systems or any system with inadequate design and development documentation above EAL 2. EAL 2 adds six new assurance components and increases the hierarchy of two components beyond those required by EAL 1 (see Exhibit 4).

Exhibit 3. EAL 1 Assurance Package

Class	Family/ Component	Requirement of Preceding EAL	New Requirement	Increased Hierarchy of Component	Principal SAR	Supporting SAR
EAL 1—Functionally Tested						
ACM Configuration Management	ACM_CAP.1 Version Numbers		X		X	
ADO Delivery and Operation	ADO_IGS.1 Installation, Generation, and Start-up Procedures		X		X	
ADV Development	ADV_FSP.1 Functional Specification		X		X	
	ADV_RCR.1 Informal Correspondence Demonstration		X			X
AGD Guidance Documents	AGD_ADM.1 Administrator Guidance		X			X
	AGD_USR.1 User Guidance		X			X
ATE Tests	ATE_IND.1 Independent Testing—conformance		X		X	

Exhibit 4. EAL 2 Assurance Package

Class	Family/ Component	Requirement of Preceding EAL	New Requirement	Increased Hierarchyof Component	Principal SAR	Supporting SAR
EAL 2—Structurally Tested						
ACM Configuration Management	ACM_CAP.2 Configuration Items			X	X	
ADO Delivery and Operation	ADO_DEL.1 Delivery Procedures		X		X	
	ADO_IGS.1 Installation, Generation, and Start-Up Procedures	X			X	
ADV Development	ADV_FSP.1 Informal Functional Specification	X			X	
	ADV_HLD.1 Descriptive High-level Design		X		X	
	ADV_RCR.1 Informal Correspondence Demonstration	X				X
AGD Guidance Documents	AGD_ADM.1 Administrator Guidance	X				X
	AGD_USR.1 User Guidance	X				X

Exhibit 4. EAL 2 Assurance Package (continued)

Class	Family/ Component	Requirement of Preceding EAL	New Requirement	Increased Hierarchy of Component	Principal SAR	Supporting SAR
ATE Tests	ATE_COV.1 Evidence of Coverage		X		X	
	ATE_FUN.1 Functional Testing		X			X
	ATE_IND.2 Independent Testing—Sample			X	X	
AVA Vulnerability Analysis	AVA_SOF.1 Strength of TOE Security Function Evaluation		X		X	
	AVA_VLA.1 Developer Vulnerability Analysis		X		X	

EAL 3, referred to as methodically tested and checked, provides a moderate level of confidence in the correct operation of the TSF.[21] EAL 3 represents a thorough investigation of the TOE and its development, starting at the design phase. Gray-box testing and evaluation are conducted against functions, interfaces, and guidance documents. EAL 3 contains at least one component from each assurance class. Four new assurance components are added and the hierarchy of three components is increased beyond those required by EAL 2 (see Exhibit 5).

EAL 4, referred to as methodically designed, tested, and reviewed, provides a moderate to high level of confidence in the correct operation of the TSF.[21] EAL 4 represents rigorous commercial development practices supplemented with proactive security engineering. EAL 4 is the first EAL to examine the security policy model. Six new assurance components are added and the hierarchy of six components is increased beyond those required by EAL 3 (see Exhibit 6).

EAL 5, referred to as semiformally designed and tested, provides a high level of confidence in the correct operation of the TSF.[21] EAL 5 represents rigorous commercial development and specialized security engineering practices and techniques. EAL 5 is appropriate in environments where resistance to attackers with a moderate attack potential is needed. Two new assurance components are added and the hierarchy of ten components is increased beyond that required by EAL 4 (see Exhibit 7).

EAL 6, referred to as semiformally verified design and tested, provides a high level of confidence in the correct operation of the TSF.[21] EAL 6 is intended to be used in high-risk environments that must protect high-value assets from attackers with a high attack potential. EAL 6 requires the use of systematic security engineering practices and techniques. No new assurance components are added; however, the hierarchy of 12 components is increased beyond those required by EAL 5 (see Exhibit 8).

EAL 7, referred to as formally verified design and tested, provides a very high level of confidence in the correct operation of the TSF.[21] EAL 7 represents complete, independent, white-box testing that employs formal methods, similar to those in use by the safety engineering community. EAL 7 is intended to be used in extremely high-risk environments that must protect high-value assets. No new assurance components are added; however, the hierarchy of eight components is increased beyond those required by EAL 6. As expected, 21 of the 25 components are at the highest possible hierarchy; ADO_IGS, ADV_LLD, and AVA_CCA are not. The ALC_FLR flaw remediation family is not included in an EAL even though it plays a major role during development, operations, and maintenance. This is due to the fact that the CEM supplement for ALC_FLR was not issued until August 2001; an updated version was released February 2002. Most likely, ALC_FLR components will be added to standard EALs in the next version of the CC (see Exhibit 9).

Some observations about EALs 5 through 7 are in order. First, evaluations at EAL 5 and above require the active involvement of the National Evaluation Authority; CCTLs cannot perform these evaluations by themselves. Second, in order to achieve EAL 5 or above (within a realistic cost and schedule), the

Exhibit 5. EAL 3 Assurance Package

Class	Family/ Component	Requirement of Preceding EAL	New Requirement	Increased Hierarchy of Component	Principal SAR	Supporting SAR
EAL 3—Methodically Tested and Checked						
ACM Configuration Management	ACM_CAP.3 Authorization Controls			X	X	
	ACM_SCP.1 TOE CM Coverage		X			X
ADO Delivery and Operation	ADO_DEL.1 Delivery Procedures	X			X	
	ADO_IGS.1 Installation, Generation, and Start-up Procedures	X			X	
ADV Development	ADV_FSP.1 Informal Functional Specification	X			X	
	ADV_HLD.2 Security Enforcing High-level Design			X	X	
	ADV_RCR.1 Informal Correspondence Demonstration	X				X

Exhibit 5. EAL 3 Assurance Package (continued)

Class	Family/ Component	Requirement of Preceding EAL	New Requirement	Increased Hierarchy of Component	Principal SAR	Supporting SAR
AGD Guidance Documents	AGD_ADM.1 Administrator Guidance	X				X
	AGD_USR.1 User Guidance	X				X
ALC Lifecycle Support	ALC_DVS.1 Identification of Security Measures		X			X
ATE Tests	ATE_COV.2 Analysis of Coverage			X	X	
	ATE_DPT.1 Testing: High-level Design		X		X	
	ATE_FUN.1 Functional Testing	X				X
	ATE_IND.2 Independent Testing—sample	X			X	
AVA Vulnerability Analysis	AVA_MSU.1 Examination of Guidance		X		X	
	AVA_SOF.1 Strength of TOE Security Function Evaluation	X			X	
	AVA_VLA.1 Developer Vulnerability Analysis	X			X	

Exhibit 6. EAL 4 Assurance Package

Class	Family/ Component	Requirement of Preceding EAL	New Requirement	Increased Hierarchy of Component	Principal SAR	Supporting SAR
ACM Configuration Management	ACM_AUT.1 Partial CM Automation		X		X	
	ACM_CAP.4 Generation Support and Acceptance Procedures			X	X	
	ACM_SCP.2 Problem Tracking CM Coverage			X	X	
ADO Delivery and Operation	ADO_DEL.2 Detection of Modification			X	X	
	ADO_IGS.1 Installation, Generation, and Start-up Procedures	X			X	
ADV Development	ADV_FSP.2 Fully Defined External Interfaces			X	X	
	ADV_HLD.2 Security Enforcing High-level Design	X			X	
	ADV_IMP.1 Subset of the Implementation of the TSF		X		X	

Exhibit 6. EAL 4 Assurance Package (continued)

Class	Family/ Component	Requirement of Preceding EAL	New Requirement	Increased Hierarchy of Component	Principal SAR	Supporting SAR
	ADV_LLD.1 Descriptive Low-level Design		X			X
	ADV_RCR.1 Informal Correspondence Demonstration	X				X
	ADV_SPM.1 Informal TOE Security Policy Model		X		X	
AGD: Guidance Documents	AGD_ADM.1 Administrator Guidance	X				X
	AGD_USR.1 User Guidance	X				X
ALC: Lifecycle Support	ALC_DVS.1 Identification of Security Measures	X				X
	ALC_LCD.1 Developer Defined Lifecycle Model		X			X
	ALC_TAT.1 Well-defined Development Tools		X			X
ATE: Tests	ATE_COV.2 Analysis of Coverage	X			X	

Exhibit 6. EAL 4 Assurance Package (continued)

Class	Family/ Component	Requirement of Preceding EAL	New Requirement	Increased Hierarchy of Component	Principal SAR	Supporting SAR
	ATE_DPT.1: Testing: High-level Design	X			X	
	ATE_FUN.1 Functional Testing	X				X
	ATE_IND.2: Independent Testing—Sample	X			X	
AVA Vulnerability Analysis	AVA_MSU.2 Validation of Analysis			X	X	
	AVA_SOF.1 Strength of TOE Security Function Evaluation	X			X	
	AVA_VLA.2 Developer Vulnerability Analysis			X	X	

Exhibit 7. EAL 5 Assurance Package

Family	Class/ Component	Requirement in Preceding EAL	New Requirement	Increased Hierarchy of Component	Principal SAR	Supporting SAR
EAL 5—Semiformally Designed and Tested						
ACM Configuration Management	ACM_AUT.1 Partial CM Automation	X			X	
	ACM_CAP.4 Generation Support and Acceptance Procedures	X			X	
	ACM_SCP.3 Development Tools CM Coverage			X	X	
ADO Delivery and Operation	ADO_DEL.2 Detection of Modification	X			X	
	ADO_IGS.1 Installation, Generation, Procedures	X			X	
ADV Development	ADV_FSP.3 Semiformal Functional Specification			X	X	
	ADV_HLD.3 Semiformal High-Level Design			X	X	
	ADV_IMP.2 Implementation of the TSF			X	X	

Exhibit 7. EAL 5 Assurance Package (continued)

Family	Class/ Component	Requirement in Preceding EAL	New Requirement	Increased Hierarchy of Component	Principal SAR	Supporting SAR
	ADV_INT.1: Modularity		X			X
	ADV_LLD.1: Descriptive Low-level Design	X				X
	ADV_RCR.2: Semiformal Correspondence Demonstration			X		X
	ADV_SPM.3: Formal TOE Security Policy Model			X	X	
AGD: Guidance Documents	AGD_ADM.1: Administrator Guidance	X				X
	AGD_USR.1: User Guidance	X				X
ALC: Lifecycle Support	ALC_DVS.1: Identification of Security Measures	X				X
	ALC_LCD.2: Standardized Lifecycle Model	X				X
	ALC_TAT.2: Compliance with Implementation Standards			X		X
ATE Tests	ATE_COV.2 Analysis of Coverage	X			X	

Exhibit 7. EAL 5 Assurance Package (continued)

Family	Class/ Component	Requirement in Preceding EAL	New Requirement	Increased Hierarchy of Component	Principal SAR	Supporting SAR
	ATE_DPT.2 Testing: low-level design			X	X	
	ATE_FUN.1 Functional Testing	X				X
	ATE_IND.2 Independent testing—sample			X	X	
AVA Vulnerability Assessment	AVA_CCA.2 Systematic Covert Channel Analysis		X		X	
	AVA_MSU.2 Validation of Analysis	X			X	
	AVA_SOF.1 Strength of TOE Security Function Evaluation	X			X	
	AVA_VLA.3 Moderately Resistant			X	X	

Exhibit 8. EAL 6 Assurance Package

Family	Class/ Component	Requirement in Preceding EAL	New Requirement	Increased Hierarchy of Component	Principal SAR	Supporting SAR
ACM Configuration Management	ACM_AUT.2 Complete CM Automation			X	X	
	ACM_CAP.5 Advanced Support			X	X	
	ACM_SCP.3 Development Tools CM Coverage	X			X	
ADO Delivery and Operation	ADO_DEL.2 Detection of Modification	X			X	
	ADO_IGS.1 Installation, Generation, and Start-up Procedures	X			X	
ADV Development	ADV_FSP.3 Semiformal Functional Specification	X			X	
	ADV_HLD.4 Semiformal High-level Explanation			X	X	
	ADV_IMP.3 Structured Implementation of the TSF			X	X	

Exhibit 8. EAL 6 Assurance Package (continued)

Family	Class/ Component	Requirement in Preceding EAL	New Requirement	Increased Hierarchy of Component	Principal SAR	Supporting SAR
	ADV_INT.2 Reduction of Complexity			X		X
	ADV_LLD.2 Semi-formal Low-Level Design			X		X
	ADV_RCR.2 Semiformal Correspondence Demonstration	X				X
	ADV_SPM.3 Formal TOE Security Policy Model	X			X	
AGD Guidance Documents	AGD_ADM.1 Administrator Guidance	X				X
	AGD_USR.1 User Guidance	X				X
ALC Lifecycle Support	ALC_DVS.2 Sufficiency of Security Measures			X		X
	ALC_LCD.2 Standardized Lifecycle Model	X				X
	ALC_TAT.3 Compliance with Implementation Standards—all parts			X		X

Exhibit 8. EAL 6 Assurance Package (continued)

Family	Class/ Component	Requirement in Preceding EAL	New Requirement	Increased Hierarchy of Component	Principal SAR	Supporting SAR
ATE Tests	ATE_COV.3 Rigorous Analysis of Coverage			X	X	
	ATE_DPT.2 Testing: low-level design	X			X	
	ATE_FUN.2 Ordered Functional Testing			X		X
	ATE_IND.2 Independent testing—sample	X			X	
AVA Vulnerability Assessment	AVA_CCA.2 Systematic Covert Channel Analysis	X			X	
	AVA_MSU.3 Analysis and Testing for Insecure States			X	X	
	AVA_SOF.1 Strength of TOE Security Function Evaluation	X			X	
	AVA_VLA.4 Highly Resistant			X	X	

Exhibit 9. EAL 7 Assurance Package

Family	Class/ Component	Requirement in Preceding EAL	New Requirement	Increased Hierarchy of Component	Principal SAR	Supporting SAR
EAL 7—Formally Verified Design and Tested						
ACM Configuration Management	ACM_AUT.2 Complete CM Automation	X			X	
	ACM_CAP.5 Advanced Support	X			X	
	ACM_SCP.3 Development Tools CM Coverage	X			X	
ADO Delivery and Operation	ADO_DEL.3 Prevention of Modification			X	X	
	ADO_IGS.1 Installation, Generation, and Start-up Procedures	X			X	
ADV Development	ADV_FSP.4 Formal Functional Specification			X	X	
	ADV_HLD.5 Formal High-level Design			X	X	
	ADV_IMP.3 Structured Implementation of the TSF	X			X	

Exhibit 9. EAL 7 Assurance Package (continued)

Family	Class/ Component	Requirement in Preceding EAL	New Requirement	Increased Hierarchy of Component	Principal SAR	Supporting SAR
	ADV_INT.3 Minimization of Complexity			X		X
	ADV_LLD.2 Semi-formal Low-Level Design	X				X
	ADV_RCR.3 Formal Correspondence Demonstration			X		X
	ADV_SPM.3 Formal TOE Security Policy Model	X			X	
AGD Guidance Documents	AGD_ADM.1 Administrator Guidance	X				X
	AGD_USR.1 User Guidance	X				X
ALC Lifecycle Support	ALC_DVS.2 Sufficiency of Security Measures	X				X
	ALC_LCD.3 Measurable Lifecycle Model			X		X
	ALC_TAT.3 Compliance with Implementation Standards—all parts	X				X
ATE Tests	ATE_COV.3 Rigorous Analysis of Coverage	X			X	

Exhibit 9. EAL 7 Assurance Package (continued)

Family	Class/ Component	Requirement in Preceding EAL	New Requirement	Increased Hierarchy of Component	Principal SAR	Supporting SAR
	ATE_DPT.3 Testing: Implementation Representation			X	X	
	ATE_FUN.2 Ordered Functional Testing	X				X
	ATE_IND.3 Independent Testing—complete			X	X	
AVA Vulnerability Assessment	AVA_CCA.2 Systematic Covert Channel Analysis	X			X	
	AVA_MSU.3 Analysis and Testing for Insecure States	X			X	
	AVA_SOF.1 Strength of TOE Security Function Evaluation	X			X	
	AVA_VLA.4 Highly Resistant	X			X	

target EAL must be determined before product design and development are initiated. The specialized security engineering techniques and practices begin during the requirements and design phases. Third, security engineering development and assurance costs start to increase dramatically from EAL 5 onward. Therefore, the selection of one of these EALs should be justified by the value of the assets to be protected and the consequences of compromise, loss, misuse, misappropriation, destruction, and so forth.

Certain terms used to define EALs or their constituent component and action elements have specific meanings within the context of the CC. For example, readers should be cognizant of the following terminology and how it is used by the CC/CEM:

- *Check:* Similar to, but less rigorous than, confirm or verify; a quick determination is made by the evaluator, perhaps requiring only a cursory analysis or perhaps no analysis at all.[21]
- *Confirm:* To review in detail in order to make an independent determination of sufficiency; the level of rigor required depends on the nature of the subject matter; applicable to evaluator actions.[21]
- *Demonstrate:* An analysis leading to a conclusion, which is less rigorous than a proof.[21]
- *Exhaustive:* Used to describe the conduct of an analysis or other activity; related to systematic but considerably stronger in that it indicates not only that a methodical approach has been taken to perform the analysis or activity according to an unambiguous plan but also that the plan that was followed is sufficient to ensure that all possible avenues have been exercised.[21]
- *Formal:* Expressed in a restricted syntax language with defined semantics based on well established mathematical concepts.[117]
- *Informal:* Expressed in a natural language.[117]
- *Justification:* An analysis leading to a conclusion but is more rigorous than a demonstration; requires significant rigor in terms of very carefully and thoroughly explaining every step of a logical argument.[21]
- *Prove:* A formal analysis in the mathematical sense that is completely rigorous in all ways.[21]
- *Semiformal:* Expressed in a restricted syntax language with defined semantics.[117]
- *Verify:* Independent evaluator actions; similar to confirm but more rigorous.[21]

Four categories of CC evaluations are conducted: PP, ST, TOE, and maintenance of assurance; they correspond to the four generic phases of a system lifecycle or acquisition process. The key players (customer, developer, sponsor, evaluator) have different roles and responsibilities in each type of evaluation.

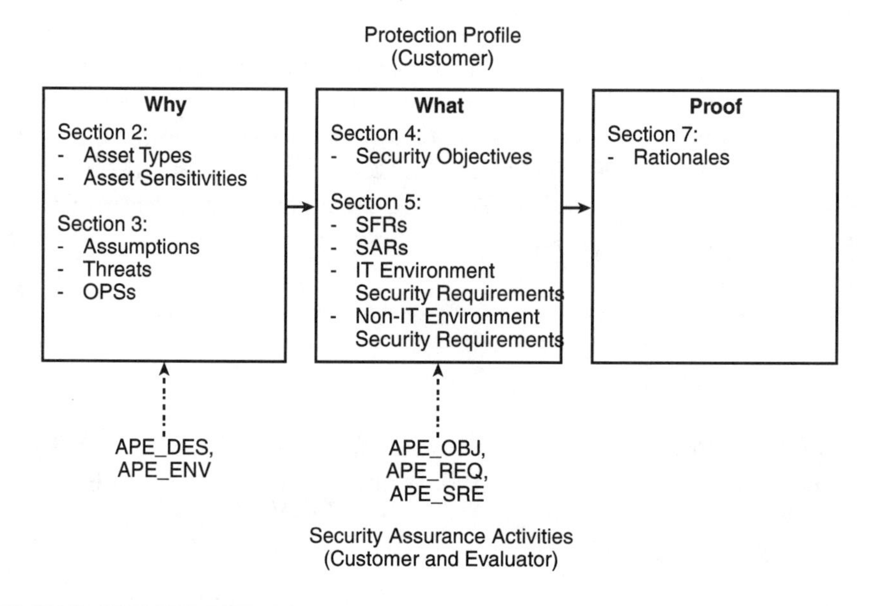

Exhibit 10. PP Evaluation

5.1.2 PP Evaluation

Exhibit 10 illustrates a PP evaluation. A customer prepares the PP. Sections 2 and 3 of the PP explain why IT security functions are necessary. Sections 4 and 5 specify what must be done to provide this protection, while Section 7 proves that the specification is correct. The customer gives the PP to an evaluator to perform a formal CC evaluation, which will ultimately lead to the issuance of a CC Certificate and an entry in the Evaluated Products List (EPL) by the National Evaluation Authority. Security assurance activities for a PP are performed by both the customer and developer; in the case of a PP, the customer equals the developer. The APE assurance class defines the PP security assurance activities. The customer is responsible for performing the developer action elements and generating the content and presentation of evidence. The evaluator is responsible for performing the evaluator action elements and evaluating the content and presentation of evidence artifacts.

Different APE families define the security assurance activities performed against the first five sections of a PP. APE_INT evaluates the PP introduction in Section 1. APE_DES evaluates the TOE description in Section 2. APE_ENV evaluates the security environment in Section 3. APE_OBJ evaluates the security objectives in Section 4. APE_REQ evaluates standard security requirements in Section 5, while APE_SRE evaluates explicit requirements. Security assurance activities that evaluate the different Rationale subsections are part of each of the above families. The APE_OBJ family will be examined to illustrate roles and responsibilities during a PP evaluation. Note that the same APE families, components, and elements apply regardless of the specified EAL.

APE_OBJ.1, the highest hierarchy component in this family, is dependent on APE_ENV.1. Security objectives are developed in response to the assumptions, threats, and organizational security policies cited in Section 3, Security Environment; hence, it is logical that the evaluation of security objectives should be dependent on the evaluation of the security environment.

APE_OBJ.1 lists two developer action elements:[21]

APE_OBJ.1.1D	*The PP developer shall provide a statement of security objectives as part of the PP.*
APE_OBJ.1.2D	*The PP developer shall provide the security objectives rationale.*

In other words, the developer is responsible for two actions: developing Section 4 of a PP that contains the security objectives and subsection 7.1 of a PP that contains the security objectives rationale.

APE_OBJ.1 lists five content and presentation of evidence elements that describe how the information in these two sections is to be provided:[21]

APE_OBJ.1.1C	*The statement of security objectives shall define the security objectives for the TOE and its environment.*
APE_OBJ.1.2C	*The security objectives for the TOE shall be clearly stated and traced back to aspects of the identified threats to be countered by the TOE and/or organizational security policies to be met by the TOE.*
APE_OBJ.1.3C	*The security objectives for the environment shall be clearly stated and traced back to aspects of identified threats not completely countered by the TOE and/or organizational security policies or assumptions not completely met by the TOE.*
APE_OBJ.1.4C	*The security objectives rationale shall demonstrate that the stated security objectives are suitable to counter the identified threats to security.*
APE_OBJ.1.5C	*The security objectives rationale shall demonstrate that the stated security objectives are suitable to cover all of the identified organizational security policies and assumptions.*

As a result, the developer is responsible for three actions related to Section 4 of a PP: (1) delineating security objectives for the TOE and the TOE environment, (2) demonstrating a direct correlation between TOE security objectives and TOE threats, and (3) demonstrating a direct correlation between security objectives for the environment and environmental assumptions, threats, and organizational security policies. In addition, the developer is responsible for two actions related to subsection 7.1 of a PP: (1) proving that the security objectives counter all identified threats, and (2) proving that the security objectives adhere to and enforce all identified assumptions and organizational security policies.

APE_OBJ.1 defines two evaluator action elements:[21]

APE_OBJ.1.1E	*The evaluator shall confirm that the information provided meets all requirements for content and presentation of evidence.*
APE_OBJ.1.2E	*The evaluator shall confirm that the statement of security objectives is complete, coherent, and internally consistent.*

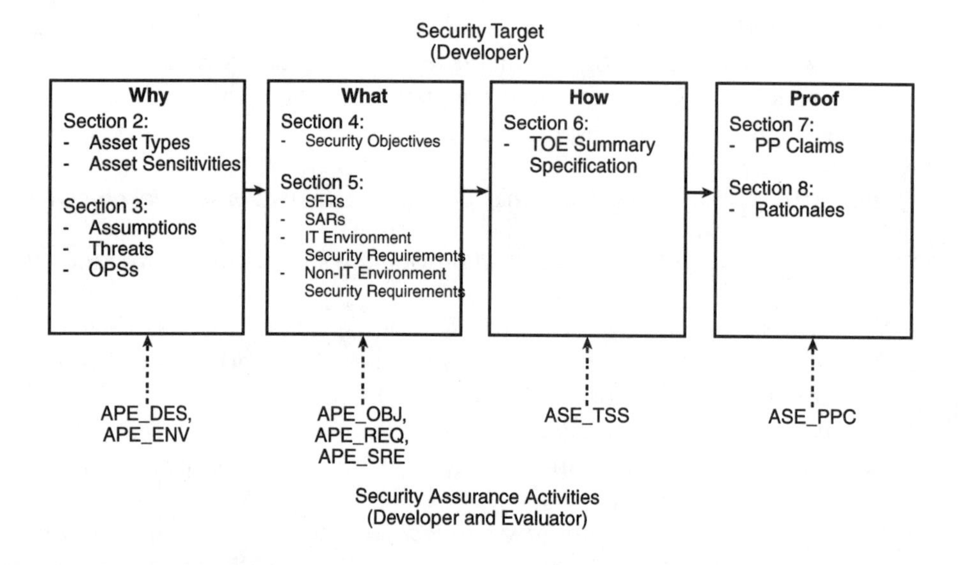

Exhibit 11. ST Evaluation

Consequently the evaluator is responsible for two actions: (1) confirming that the information required by the five content and presentation of evidence elements has been supplied, and (2) confirming that the security objectives as stated are complete, coherent, and internally consistent. Evaluators may use any evaluation techniques they deem appropriate to perform these actions. (Discussion of the CEM, which follows, explains how feedback from an evaluation is disseminated, when, and to whom.)

5.1.3 ST Evaluation

Exhibit 11 illustrates an ST evaluation. A developer prepares an ST in response to a customer's PP.* Sections 2 and 3 explain why IT security functions are needed. Sections 4 and 5 specify what must be done to provide this protection. Section 6 describes how security functions, mechanisms, and assurance measures will provide the level of protection required. Sections 7 and 8 prove that the security solution is correct, coherent, complete, and consistent internally with the referenced PP. The sponsor gives the ST to an evaluator to perform a formal CC evaluation, which will ultimately lead to issuance of a CC Certificate and an entry in the EPL by the National Evaluation Authority. Security assurance activities for an ST are performed by both the developer and evaluator. The ASE assurance class defines the ST security assurance activities. The developer is responsible for performing the developer action elements and generating the content and presentation of evidence. The eval-

* The preferred sequence of events is for the customer to write a PP and the developer to respond with an ST. Should a PP not exist, however, a developer may create an ST in the absence of a PP. In this case, Sections 1 through 5 must be approved by the customer prior to developing Sections 6 through 8 of the ST or the TOE itself.

uator is responsible for performing the evaluator action elements and evaluating the content and presentation evidence artifacts. An ST may be submitted for evaluation prior to the TOE evaluation or concurrent with it. Common sense dictates that an ST be submitted for evaluation before the TOE so that design defects can be corrected prior to construction.

Different ASE families define the security assurance activities performed against the eight sections of an ST. ASE_INT evaluates the ST introduction in Section 1. ASE_DES evaluates the TOE description in Section 2. ASE_ENV evaluates the security environment in Section 3. ASE_OBJ evaluates the security objectives in Section 4. ASE_REQ evaluates standard security requirements in Section 5, while ASE_SRE evaluates explicit requirements. ASE_TSS evaluates the TSS in Section 6, and ASE_PPC evaluates the PP Claims in Section 7. Security assurance activities that evaluate the different Rationale subsections are part of each of the above families. The ASE_SRE family will be examined to illustrate roles and responsibilities during an ST evaluation. Note that the same ASE families, components, and elements apply regardless of the specified EAL.

ASE_SRE.1, the highest hierarchy, is dependent on ASE_REQ.1. In essence, the evaluation of explicit requirements is dependent on the evaluation of standard requirements. ASE_SRE.1 lists two developer action elements:[21]

ASE_SRE.1.1D	*The developer shall provide a statement of IT security requirements as part of the ST.*
ASE_SRE.1.2D	*The developer shall provide the security requirements rationale.*

These two developer action elements are the same as those for ASE_REQ; the developer is responsible for generating security requirements in Section 5 of the ST and proving that they are correct in subsection 8.2 of the ST.

ASE_SRE.1 lists seven content and presentation of evidence elements that describe how the information in these two sections is to be provided:[21]

ASE_SRE.1.1C	*All TOE security requirements that are explicitly stated without reference to ISO/IEC 15408 shall be identified.*
ASE_SRE.1.2C	*All security requirements for the IT environment that are explicitly stated without reference to ISO/IEC 15408 shall be identified.*
ASE_SRE.1.3C	*The evidence shall justify why the security requirements had to be explicitly stated.*
ASE_SRE.1.4C	*The explicitly stated IT security requirements shall use the ISO/IEC 15408 requirements components, families, and classes as a model for presentation.*
ASE_SRE.1.5C	*The explicitly stated IT security requirements shall be measurable and state objective evaluation requirements such that compliance or noncompliance of a TOE can be determined and systematically demonstrated.*
ASE_SRE.1.6C	*The explicitly stated IT security requirements shall be clearly and unambiguously expressed.*
ASE_SRE.1.7C	*The security requirements rationale shall demonstrate that the assurance requirements are applicable and appropriate to support any explicitly stated TOE security functional requirements.*

As a result, the developer has six actions regarding explicit requirements in Section 5 and one action in regard to subsection 8.2. Explicit security requirements for the TOE and the TOE environment have to be identified. The use of explicit requirements, rather than standard requirements, has to be justified. Explicit requirements have to be expressed using the standard CC syntax and notation. Explicit requirements have to be expressed so that their achievement can be objectively measured. Explicit requirements have to be stated unambiguously. Evaluation criteria for explicit requirements have to be defined and proven to be appropriate and sufficient.

ASE_SRE.1 defines two evaluator action elements:[21]

ASE_SRE.1.1E	*The evaluator shall confirm that the information provided meets all requirements for content and presentation of evidence.*
ASE_SRE.1.2E	*The evaluator shall determine that all of the dependencies of the explicitly stated IT security requirements have been identified.*

Consequently, the evaluator is responsible for confirming that the information required by the seven content and presentation of evidence elements has been supplied and that all dependencies of explicitly stated requirements have been highlighted. The latter is particularly important because explicitly stated requirements may have dependencies on standard or other explicit requirements, both of which are necessary if the TSF is to function correctly. Again, evaluators may use any evaluation techniques they deem appropriate to complete these actions.

A TOE evaluation ensures, among other things, that the TOE security policy (TSP) is accurately and adequately enforced across all TOE resources.[20] The TSP is a set of rules that regulate how assets are managed, protected, and distributed within a TOE.[19] TSP is generally composed of multiple security policies, each of which: (1) has its own scope of control, and (2) defines subjects, objects, and permitted operations. While an ST undergoes a formal CCTL evaluation and may pass or fail, CC Certificates are not issued for STs; CC Certificates are only issued for PPs and TOEs (refer to subsection 5.4).

5.1.4 TOE Evaluation

Exhibit 12 illustrates a TOE evaluation. A developer prepares the "as-built" product or system, including all hardware, software, firmware, and documentation, and submits it to the sponsor. The sponsor turns the TOE over to an evaluator (CCTL) to perform a formal CC evaluation, which will ultimately lead to the issuance of a CC Certificate and an entry in the EPL by the National Evaluation Authority. Security assurance activities for a TOE are performed by both the developer and the evaluator. Several security assurance classes are involved: ACM, ADO, ADV, AGD, ALC, ATE, and AVA. The specific activities performed are determined by the EAL and SARs specified in the PP. The developer is responsible for performing the developer action elements and generating the content and presentation of evidence criteria. The evaluator is responsible for performing the evaluator action elements and evaluating the content and presentation evidence artifacts.

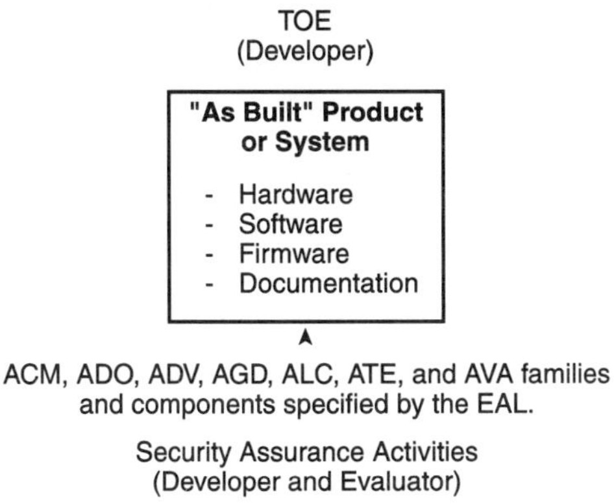

TOE
(Developer)

"As Built" Product or System
- Hardware - Software - Firmware - Documentation

ACM, ADO, ADV, AGD, ALC, ATE, and AVA families
and components specified by the EAL.

Security Assurance Activities
(Developer and Evaluator)

Exhibit 12. TOE Evaluation

The sponsor should establish a relationship with an evaluator (CCTL) prior to TOE construction. To optimize schedule performance, all parties involved (customer, sponsor, developer, and evaluator) should understand their roles and responsibilities before a formal evaluation begins. It is particularly important for developers to understand all the CC artifacts and evidence they are responsible for generating. Likewise, the customer or sponsor should ensure that their contract with the developer specifies all the necessary CC artifacts and evidence as contract deliverables (CDRLs). Note that the CC specifies the content of the artifacts and evidence, not the format. The customer or sponsor is responsible for defining the preferred format. Lack of this evidence and poor-quality evidence are the most common reasons for delays in completing a TOE evaluation. The earlier a CCTL is involved, the better; a CCTL can begin a TOE evaluation as soon as the design is (essentially) solidified. The time to complete an evaluation is much longer if the CCTL is not involved, until after the TOE is constructed. All defects must be corrected before a product or system can be certified, and it is easier and less expensive to detect defects during the design phase than after development.

The specified EAL defines what security assurance activities are performed during a TOE evaluation. The families, components, and component hierarchies invoked vary by EAL. To illustrate, EAL 3 is the first EAL to contain security assurance activities from at least one family in each assurance class. EAL 5 is the first EAL to contain security assurance activities from every assurance class and family, except ALC_FLR. The ADO_IGS family will be examined to illustrate roles and responsibilities during a TOE evaluation.

ADO_IGS.2, the highest in the hierarchy, is dependent on AGD_ADM.1. This means that installation, generation, and start-up procedures are dependent on administrator guidance. ADO_IGS.2 defines one developer action element:[21]

ADO_IGS.2.1D *The developer shall document procedures necessary for the secure installation, generation, and start-up of the TOE.*

The developer is responsible for preparing installation, generation, and start-up procedures that will initiate the TOE in a secure manner. This information may be provided as a stand-alone document or as part of another document in accordance with the sponsor's preference.

ADO_IGS.2 defines two content and presentation of evidence elements:[21]

ADO_IGS.2.1C	*The documentation shall describe the steps necessary for secure installation, generation, and start-up of the TOE.*
ADO_IGS.2.2C	*The documentation shall describe procedures capable of creating a log containing the generation options used to generate the TOE in such a way that it is possible to determine exactly how and when the TOE was generated.*

These two elements further refine the information content that the developer is responsible for providing. The first piece of evidence describes in detail the steps to follow to generate the TOE, while the second captures the exact options invoked during one generation exercise.

ADO_IGS.2 defines two evaluator action elements:[21]

ADO_IGS.2.1E	*The evaluator shall confirm that the information provided meets all requirements for content and presentation of evidence.*
ADO_IGS.2.2E	*The evaluator shall determine that the installation, generation, and start-up procedures result in a secure configuration.*

The evaluator is responsible for (1) confirming that all of the information required by the content and presentation of evidence elements has been provided, and (2) determining that the procedures, if followed correctly, generate a secure instance of the TOE. Evaluators may use any evaluation technique(s) they deem appropriate to perform these two action elements.

5.1.5 Maintenance of Assurance Evaluation

Passing the initial CC certification of a product or system is a major accomplishment, but this is not the end; rather, it is the beginning of a series of ongoing security assurance activities. It is during the operations and maintenance phase that the real effectiveness, robustness, and resilience of the TSF are proven. Correct performance in the operational environment is the ultimate goal; this is the reason security assurance activities are specified and undertaken in the first place.

Exhibit 13 illustrates a Maintenance of Assurance evaluation. The organization responsible for operating and maintaining the "in-service" product or system, including all hardware, software, firmware, and documentation, makes it available to the sponsor. The organization responsible for operating and maintaining the "in-service" product or system could be the developer, customer, or a third-party system integrator. The sponsor turns the maintenance of assurance evidence and TOE over to an evaluator (CCTL) to perform a formal CC audit to ensure that the EAL to which the product or system was certified is being maintained through the performance of security assurance activities. Maintenance of assurance activities for a TOE are performed by both the organization responsible for operating and maintaining the "in-

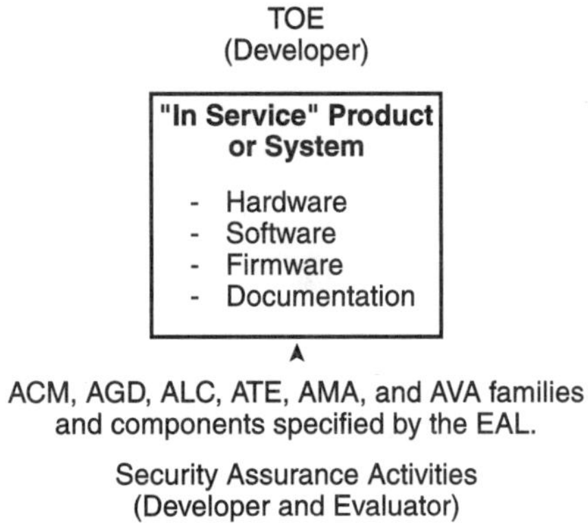

TOE
(Developer)

"In Service" Product
or System

- Hardware
- Software
- Firmware
- Documentation

ACM, AGD, ALC, ATE, AMA, and AVA families
and components specified by the EAL.

Security Assurance Activities
(Developer and Evaluator)

Exhibit 13. Maintenance of Assurance Evaluation

service" product or system and the evaluator. Several security assurance classes are involved: ACM, AGD, ALC, ATE, AMA, and AVA. The specific activities performed are determined by the EAL and SARs specified in the PP. The organization responsible for operating and maintaining the "in-service" product or system is responsible for performing the developer action elements and generating the content and presentation of evidence criteria. The evaluator is responsible for performing the evaluator action elements and evaluating the content and presentation evidence artifacts.

Maintenance of assurance requirements must be defined in a PP before the TOE is constructed; it is difficult to retrofit maintenance of assurance without significant cost and schedule impact. Responsibilities for performing maintenance of assurance activities (customer, developer, or third-party system integrator) must be delineated during development of the PP as well. For example, the customer plays a key role on the working group that evaluates results of security impact analyses of proposed enhancements, changes, or fixes to a system or product. Successful performance of maintenance of assurance activities is dependent on prior planning and coordination. The sponsor should establish an ongoing relationship with the evaluator that continues after initial certification. Most National Evaluation Authorities require the evaluator to conduct annual audits, at a minimum, to ensure that maintenance of assurance activities are being performed correctly. These annual audits also provide insight into when the threshold has been crossed from maintenance of assurance activities to the need for a complete recertification evaluation (this topic is discussed further below and in Chapter 6).

Maintenance of assurance activities are derived from several assurance classes, including ACM, AGD, ALC, ATE, AMA, and AVA. The EAL and SARs specified in the PP, including any augmentations and extensions, determine which activities are performed. The AMA_SIA family will be examined to illustrate roles and responsibilities during a maintenance of assurance evaluation.

AMA_SIA.1, the lowest hierarchy component in this family, is dependent on AMA_CAT.1. The component categorization report ranks TOE components by their importance to TOE security; hence, it is essential input to any security impact analysis. AMA_SIA.1 defines one developer action element:[21]

AMA_SIA.1.1D	*The developer security analyst shall, for the current version of the TOE, provide a security impact analysis that covers all changes affecting the TOE as compared with the certified version.*

The developer is required to appoint a security analyst, who has an independent reporting channel from the development team, to assess the security impact of all proposed changes to the TOE, whether as a result of corrective, adaptive, or preventive maintenance. The customer may also want to have an independent third party perform security impact analyses to remove any potential or perceived conflicts of interest, especially if the developer is the organization responsible for performing post-evaluation operation and maintenance of the TSF.

Note that the standard differentiates between the certified TOE and the current version of the TOE:[21]

- *Certified TOE* — Version of the TOE that was evaluated, awarded a CC Certificate, and is listed in the evaluated product's list of a National Evaluation Authority.
- *Current version of TOE* — Version of the TOE that differs in some respect from the certified version, such as (1) a new release of the TOE, (2) a certified version with patches to correct subsequently discovered bugs, and (3) the same basic version of the TOE but on a different hardware or software platform.

The first scenario applies to adaptive maintenance, the second scenario applies to corrective and preventive maintenance, and the third scenario applies to a change in the intended TOE operational environment.

AMA_SIA.1 defines seven content and presentation of evidence elements:[21]

AMA_SIA.1.1C	*The security impact analysis shall identify the certified TOE from which the current version of the TOE was derived.*
AMA_SIA.1.2C	*The security impact analysis shall identify all new and modified TOE components that are categorized as TSP-enforcing.*
AMA_SIA.1.3C	*The security impact analysis shall, for each change affecting the security target or TSF representations, briefly describe the change and any effects it has on lower representation levels.*
AMA_SIA.1.4C	*The security impact analysis shall, for each change affecting the security target or TSF representations, identify all IT security functions and all TOE components categorized as TSP-enforcing that are affected by the change.*

AMA_SIA.1.5C	*The security impact analysis shall, for each change which results in a modification of the implementation representation of the TSF or the IT environment, identify the test evidence that show, to the required level of assurance, that the TSF continues to be correctly implemented following the change.*
AMA_SIA.1.6C	*The security impact analysis shall, for each applicable assurance requirement in the configuration management (ACM), lifecycle support (ALC), delivery and operation (ADO), and guidance documents (AGD) assurance classes, identify any evaluation deliverables that have changed, and provide a brief description of each change and its impact on assurance.*
AMA_SIA.1.7C	*The security impact analysis shall, for each applicable assurance requirement in the vulnerability assessment (AVA) assurance class, identify which evaluation deliverables have changed and which have not, and give reasons for the decision taken as to whether or not to update the deliverable.*

In summary, the developer is responsible for identifying what has changed, how it was changed, and how the changes impact the TSF. New and modified TSP-enforcing functions and mechanisms must be identified. The ripple effect of the changes throughout the TSF design must be highlighted. Test evidence must be provided that proves that the change does not cause the TSF to operate incorrectly. The impact of the changes on the results of other security assurance measures (ACM, ALC, ADO, ALC, and AVA), including the associated supporting documentation, must be described. While not required by the CC, the customer may want to have an independent third party generate this evidence as well.

AMA_SIA.1 defines two evaluator action elements:[21]

AMA_SIA.1.1E	*The evaluator shall confirm that the information provided meets all requirements for content and presentation of evidence.*
AMA_SIA.1.2E	*The evaluator shall check, by sampling, that the security impact analysis documents changes to an appropriate level of detail, together with appropriate justifications that assurance has been maintained in the current version of the TOE.*

In summary, the evaluator is responsible for confirming that the security impact analysis was conducted in a thorough manner and that the results provided are accurate and complete.

5.2 Common Evaluation Methodology (CEM)

The ISO/IEC defines the assurance classes, families, components, developer action elements, content and presentation of evidence criteria, evaluator action elements, and standard EALs. The CEM builds upon this foundation by elaborating roles, responsibilities, activities, subactivities, actions, and work units to be fulfilled.

The CEM is developed and maintained by the Common Evaluation Methodology Editing Board (CEMEB), under the CC Implementation Management Board (CCIMB). At present, the CEM consists of two parts and a supplement, for a total of 424 pages:

- CEM-97/017, Common Evaluation Methodology for Information Technology Security, Part 1: Introduction and general model, (draft) version 0.6, 11 January 1997
- CEM-99/045, Common Evaluation Methodology for Information Technology Security, Part 2: Evaluation Methodology, version 1.0, August 1999
- CEM-2001/0015R, Common Evaluation Methodology for Information Technology Security, Part 2: Evaluation Methodology, Supplement: ALC_FLR Flaw Remediation, version 1.1, February 2002

Part 1 of the CEM establishes the universal principles or goals that apply to all CC/CEM evaluations:

- *Appropriateness:* The evaluation activities employed in achieving an intended level of assurance shall be appropriate.
- *Impartiality:* All evaluations shall be free from bias.
- *Objectivity:* Evaluation results shall be obtained with a minimum of subjective judgment or opinion.
- *Repeatability and reproducibility:* The repeated evaluation of the same TOE or PP to the same requirements with the same evaluation evidence shall yield the same results.
- *Soundness of results:* The evaluation results shall be complete and technically correct.

Part 1 also establishes the basic roles and responsibilities during an evaluation. Four roles are defined: for the sponsor, developer, evaluator, and validator. The sponsor is the entity that wants the evaluation performed; this may be the customer or developer organization. The sponsor is responsible for initiating a contractual relationship with the evaluator and delivering all the CC artifacts required for an evaluation. The developer is the entity that generates the CC artifacts that are being subjected to an evaluation. Developers are responsible for providing support to the evaluator, on an as-requested basis, performing the developer action elements specified by the EAL in the PP and maintaining the associated content and presentation of evidence. The evaluator is an accredited CCTL selected by the sponsor. Different organizations may perform the role of evaluator during initial certification and maintenance of assurance. The evaluator receives the CC artifacts to be evaluated from the sponsor. The evaluator is responsible for performing the evaluator action elements specified by the EAL in the PP, including any augmentations or extensions. During the evaluation, the evaluator is expected to uphold the universal principles and, if needed, request clarification from the developer or validator. Interim and final results of the evaluation are documented by the evaluator. The evaluator can yield three possible verdicts:

1. *Pass* — All constituent evaluator action elements are met and all requirements are met. The developer has satisfied all SFRs, SARs, developer action elements, and content and presentation of evidence criteria.
2. *Inconclusive* — Any of the constituent evaluator actions are deemed incomplete. The evaluator cannot successfully complete all of the required evaluator action elements due to inadequate, incomplete, or ambiguous content and presentation of evidence.
3. *Fail* — All constituent evaluator actions are complete and it is determined that one or more of the requirements are not met. The evaluator has completed the evaluation; however, defects or anomalies were discovered in the implementation of SFRs, the performance of developer actions, and/or the content and presentation of evidence.

The validator is the National Evaluation Authority. The validator is responsible, as a participant in the CCRA, for defining their national evaluation scheme, accrediting CCTLs, and monitoring their performance. Validators review the formal evaluation results from the CCTLs and either approve or reject the results. If approved, validators issue a CC Certificate and make an entry in the EPL.

Like ISO/IEC 15408, the CEM defines and uses terms within a certain context. It is important for the reader to be aware of this terminology and its usage.[25]

- *Action* — Explicitly described CC evaluator action element or one derived from a specified developer action element
- *Activity* — Application of a CC assurance class
- *Evaluation Technical Report (ETR)* — Report that documents the overall verdict and its justification; produced by the evaluator and submitted to the validator and sponsor
- *Observation Report (OR)* — Report written by an evaluator requesting clarification or identifying a problem during the evaluation
- *Subactivity* — Application of a CC assurance component
- *Subtask* — CEM-specific evaluation work that is not derived directly from CC requirements
- *Task* — CEM-specific evaluation work that is not derived directly from CC requirements
- *Verdict* — Pass, fail, or inconclusive statement issued by an evaluator with respect to a CC evaluator action element, assurance component, or class
- *Work unit* — Smallest unit of an evaluation action; derived from an evaluator action element or a content and presentation of evidence element

These terms are reflected in the CEM syntax and notation. For example, consider:

3:ADO_IGS.1-2

The "3" indicates an EAL 3 evaluation, "ADO" indicates the evaluation activity, "IGS.1" indicates the evaluation subactivity, and the "-2" indicates that this is the second work unit in the ADO_IGS.1 subactivity. Work unit numbers may change from one EAL to another, because of increasing scope, depth, and rigor.

At a high level, evaluation tasks can be grouped into two categories: input tasks (management of evaluation evidence) and output tasks (report generation). These tasks are an example of how the CEM adds evaluation process details beyond those described in the CC.

The purpose of the input task is to ensure that the evaluator has a complete and current set of evidence. The sponsor is responsible for delivering all the evidence to the evaluator. The CEM recommends supplying an evidence index that provides an inventory with current version numbers. The evaluator is responsible for three input subtasks:

1. Maintaining the exact configuration of the evidence as delivered and protecting it from accidental or intentional modification or loss
2. Protecting the confidentiality of evidence consistent with the sensitivity determined by the sponsor
3. Disposing of the evaluation evidence afterward in a manner that is mutually agreeable to the sponsor

The purpose of the output task is to document the observations and conclusions resulting from the conduct of an evaluation. The CEM specifies the content and format of the information to be captured to ensure the consistency and repeatability of evaluations. National evaluation schemes may require additional information, but not less; the CEM defines the minimum acceptable informational content of evaluation reports.

The CEM defines two output substasks: the Observation Report (OR) and the Evaluation Technical Report (ETR). ORs are used for two purposes during the conduct of an evaluation: (1) to issue a request for clarification, and (2) to identify a problem. ORs are considered formal evaluation documents. They are written by an evaluator and sent simultaneously to the sponsor and the National Evaluation Authority.

Exhibit 14 illustrates the contents of an OR. The same format is used, regardless of whether the evaluation is for a PP, ST, or TOE. Section 1 provides identification information, such as the PP or ST identification section supplemented with version numbers, release dates, hardware/software platforms, and configuration options. Section 2 identifies the task or subactivity in which the problem occurred or an observation was made that requires clarification. Section 3 details the problem or request for clarification — for example, what went wrong; missing, ambiguous, conflicting, or incorrect evidence; or interim

Exhibit 14. Content of an Observation Report (OR)

1. PP, ST, or TOE Identifier
2. Task/Sub-activity in which encountered
3. Request for Clarification or Problem
4. Severity Assessment
5. Resolution Agent
6. Recommended Resolution Timetable
7. Impact Assessment

results that could lead to a fail verdict.[25] Section 4 states the severity of the OR in regard to the overall evaluation. Section 5 assigns responsibility for responding to an OR. Note that an OR could be generated that results in a Request for Interpretation (RI) to the National Evaluation Authority. Section 6 documents the evaluator's recommended timetable for resolving the OR, while Section 7 describes the impact on the overall evaluation should the OR not be resolved. Multiple ORs may be generated throughout the conduct of an evaluation. ORs and the resolution evidence are kept under configuration management control.

The purpose of an ETR is to document and justify the evaluation results, in particular, the rationale for the verdict. An ETR is written by an evaluator. The primary audience of an ETR is the sponsor and National Evaluation Authority; the secondary audience is the developer and potential customers who are trying to determine if the TOE is appropriate for their environment.

Exhibit 15 illustrates the contents of an ETR. This format is used for an ST or TOE evaluation; Section 2 is omitted for a PP evaluation. Section 1, Introduction, is required by the CEM to contain seven key pieces of information:[25]

1. National evaluation scheme under which the evaluation was conducted
2. ETR configuration control identifier
3. PP, ST, and/or TOE configuration control identifiers
4. Referenced PP to which the ST and/or TOE claim conformance
5. Developer's identity
6. Sponsor's identity
7. Evaluator's identity

Section 2 describes the high-level architecture of the TOE, such as relationships between functional packages, component TOEs, and the composite TOE. At the same time, the scope of the evaluated configuration is clarified, including specifying which functional packages and component TOEs were and were not included in the evaluation. Section 3 documents the methods, techniques, and tools that were used to conduct the evaluation. Assumptions and constraints that may have affected the evaluation results are also highlighted. Section 4 captures the detailed results of an evaluation. Results are provided by activity, with the exception of ATE and AVA, which are presented by work unit. The results consist of a verdict and a detailed justification that supports

Exhibit 15. Content of an Evaluation Technical Report (ETR)

1. Introduction
2. Architectural Description of the TOE (for PP evaluations)
3. Evaluation
4. Evaluation Results
5. Conclusions and Recommendations
6. Evaluation Evidence Index
7. Glossary of Acronyms and Terms
8. Chronology and Resolution of ORs

that verdict. The evaluator's conclusions and recommendations for the National Evaluation Authority are documented in Section 5. The evaluation evidence index is included in Section 6; this information also clarifies the scope of the evaluated configuration. Section 7 contains a glossary of acronyms and terms used in the ETR. Finally, Section 8 cites all ORs that were generated during the conduct of the evaluation and their formal resolution.

The following examples illustrate how the CEM builds upon the foundation established by ISO/IEC 15408-3. The first example is for a PP evaluation, the second for a TOE evaluation. For each subactivity, the CEM states the evaluation objectives, provides application notes, cites the mandatory evidence that is an input to the subactivity, defines evaluator work units, and provides associated guidance.

APE_SRE.1 defines two evaluator action elements:[21]

APE_SRE.1.1E	*The evaluator shall confirm that the information provided meets all requirements for content and presentation of evidence.*
APE_SRE.1.2E	*The evaluator shall determine that all of the dependencies of the explicitly stated IT security requirements have been identified.*

The CEM defines seven work units for APE_SRE.1.1E and one work unit for APE_SRE.1.2E:[25]

APE_SRE.1-1	*The evaluator shall check that the statement of the IT security requirements identifies all TOE security requirements that are explicitly stated without reference to the CC.*
APE_SRE.1-2	*The evaluator shall check that the statement of IT security requirements identifies all security requirements for the IT environment that are explicitly stated without reference to the CC.*
APE_SRE.1-3	*The evaluator shall examine the security requirements rationale to determine that it appropriately justifies why each explicitly stated IT security requirement had to be explicitly stated.*
APE_SRE.1-4	*The evaluator shall examine each explicitly stated IT security requirement to determine that the requirement uses the CC requirements components, families, and classes as a model for presentation.*

APE_SRE.1-5	*The evaluator shall examine each explicitly stated IT security requirement to determine that it is measurable and states objective evaluation requirements, such that compliance or noncompliance of a TOE can be determined and systematically demonstrated.*
APE_SRE.1-6	*The evaluator shall examine each explicitly stated IT security requirement to determine that it is clearly and unambiguously expressed.*
APE_SRE.1-7	*The evaluator shall examine the security requirements rationale to determine that it demonstrates that the assurance requirements are applicable and appropriate to support any explicitly stated TOE security functional requirements.*
APE_SRE.1-8	*The evaluator shall examine the statement of IT security requirements to determine that all of the dependencies of any explicitly stated IT security requirements have been identified.*

The first seven work units correspond to the seven content and presentation of evidence criteria in ISO/IEC 15408-3. The eighth work unit evaluates the completeness of the explicitly stated requirements. In addition, the CEM provides evaluator guidance for almost each work unit. For example, the following guidance is provided for APE_SRE.1-7:[25]

> The evaluator determines whether application of the specified assurance requirements will yield a meaningful evaluation result for each explicitly stated security functional requirement, or whether other assurance requirements should have been specified. For example, an explicitly stated functional requirement may imply the need for particular documentary evidence (such as TSP model), depth of testing, or analysis (such as strength of TOE security function analysis or covert channel analysis).

AVA_VLA.1 defines two evaluator action elements:[21]

AVA_VLA.1.1E	*The evaluator shall confirm that the information provided meets all requirements for content and presentation of evidence.*
AVA_VLA.1.2E	*The evaluator shall conduct penetration testing, building on the developer vulnerability analysis, to ensure obvious vulnerabilities have been addressed.*

The CEM defines three work units for AVA_VLA.1.1E and seven work units for AVA_VLA.1.2E when EAL 3 is specified:[25]

3:AVA_VLA.1-1	*The evaluator shall examine the developer's vulnerability analysis to determine that the search for obvious vulnerabilities has considered all relevant information.*
3:AVA_VLA.1-2	*The evaluator shall examine the developer's vulnerability analysis to determine that each obvious vulnerability is described and that a rationale is given for why it is not exploitable in the intended environment for the TOE.*

3:AVA_VLA.1-3 *The evaluator shall examine the developer's vulnerability analysis to determine that it is consistent with the ST and the guidance.*

3:AVA_VLA.1-4 *The evaluator shall devise penetration tests, building on the developer vulnerability analysis.*

3:AVA_VLA.1-5 *The evaluator shall produce penetration test documentation for the tests that build upon the developer vulnerability analysis, in sufficient detail to enable the tests to be repeatable. The test documentation shall include:*

a) *identification of the obvious vulnerability the TOE is being tested for;*
b) *instructions to connect and set up all required test equipment as required to conduct the penetration test;*
c) *instructions to establish all penetration test prerequisite initial conditions;*
d) *instructions to stimulate the TSF;*
e) *instructions for observing the behavior of the TSF;*
f) *descriptions of all expected results and the necessary analysis to be performed on the observed behavior for comparison against expected results;*
g) *instructions to conclude the test and establish the necessary post-test state for the TOE.*

3:AVA_VLA.1-6 *The evaluator shall conduct penetration testing, building on the developer vulnerability analysis.*

3:AVA_VLA.1-7 *The evaluator shall record the actual results of the penetration tests.*

3:AVA_VLA.1-8 *The evaluator shall examine the results of all penetration testing and the conclusions of all vulnerability analysis to determine that the TOE, in its intended environment, has no exploitable obvious vulnerabilities.*

3:AVA_VLA.1-9 *The evaluator shall report in the ETR the evaluator penetration testing effort, outlining the testing approach, configuration, depth and results.*

3:AVA_VLA.1-10 *The evaluator shall report in the ETR all exploitable vulnerabilities and residual vulnerabilities, detailing for each:*

a) *Its source*
b) *The implicated security functions(s), objective(s) not met, OSPs contravened, and threats realized*
c) *A description*
d) *Whether it is exploitable in its intended environment or not*
e) *Identification of evaluation party*

In this example, one content and presentation of evidence criterion translates into ten evaluator work units. Again, evaluator guidance is provided for most work units.

The CEM Part 2 supplement defines the methodology for applying the CC assurance requirements of the ALC_FLR Flaw Remediation family; these requirements were not addressed in the original issue of Part 2. This supplement includes Final Interpretations 062 and 092 and supersedes Final Interpretation 094. The next version of Part 2 of the CEM will incorporate this supplement. The description of the ALC_FLR family in ISO/IEC 15408-3 will be updated at the same time. At present, ALC_FLR components are not assigned to any EAL but may be added as extensions when developing a PP or ST.

5.3 National Evaluation Schemes

A scheme is defined as:[25]

> a set of rules, established by a National Evaluation Authority, defining the evaluation environment, including criteria and methodology required to conduct IT security evaluations.

Expressed another way, a national evaluation scheme:[19]

> ...sets the standards, monitors the quality of evaluations, administers the regulations to which evaluation facilities and evaluators must conform.

CCRA participants who issue CC Certificates must define a national evaluation scheme. CCRA participants who consume CC Certificates are not required to do so.

ISO/IEC 15408-3 defines the assurance classes, families, components, developer action elements, content and presentation of evidence criteria, evaluator action elements, and standard EALs. The CEM defines roles, responsibilities, activities, subactivities, actions, and work units to be fulfilled. Schemes promulgated by each National Evaluation Authority build upon this foundation by enumerating details about how the evaluation process is conducted and managed. National evaluation schemes may add to CEM requirements, but they may not subtract from them. Annex B.29 of Part 2 of the CEM lists 20 potential items that National Evaluation Authorities may choose to specify, such as the language in which evaluation evidence is submitted and the use of scheme identifiers, logos, and trademarks.

In the United States, the National Information Assurance Partnership (NIAP®) is responsible for defining the national evaluation scheme. At the time of writing, five of six planned scheme publications have been issued. The sixth is planned for late December 2002 (see Chapter 6). In addition, several templates have been developed for CC artifacts. The NIAP® scheme is referred to as the Common Criteria Evaluation and Validation Scheme (CCEVS).

CCEVS Publication 1, *Organization, Management and Concept of Operations*, version 2.0, was issued in May 1999. This document provides a general overview of the purpose and conduct of the CCEVS and the roles and responsibilities of NIAP®, the National Voluntary Laboratory Accreditation Program (NVLAP®), CCTLs, and sponsors. The content and format of CC Certificates issued by NIAP® are illustrated, along with the NIAP® logo.

CCEVS Publication 2, *Validation Body Standard Operating Procedures*, draft version 1.5, was issued in May 2000. This document provides a detailed description of the roles, responsibilities, organization, management, and operation of NIAP®.

CCEVS Publication 3, *Guidance to Validators of IT Security Evaluations*, version 1.0, was issued in February 2002. This document provides a detailed discussion of the CCEVS verification and validation processes and the associated roles and responsibilities of NIAP®, NVLAP®, and the CCTLs. In addition, guidance is provided for CEM work units and formats are specified for evaluation records.

CCEVS Publication 4, *Guidance to CCEVS Accredited CCTLs*, draft version 1.0, was issued in March 2001. This document details the procedure for becoming an accredited CCTL and CCTL pre-evaluation, evaluation, and post-evaluation responsibilities. Templates are provided, such as a CCEVS evaluation work plan template and a proposed agenda for the kick-off meeting of NIAP®, the CCTL, and sponsor.

CCEVS Publication 5, *Guidance to Sponsors of IT Security Evaluations*, draft version 1.0, was issued in August 2000. This document details the responsibilities of sponsors during the pre-evaluation, evaluation, and post-evaluation phases.

CCEVS Publication 6, *Certificate Maintenance Program (CMP)*, is scheduled for publication in late December 2002. This document will provide guidance to sponsors, developers, and evaluators on how to maintain a CC Certificate and the associated EAL during the operations and maintenance phase.

The CCEVS consists of four major phases each of which is discussed below:

1. Preparation
2. Conduct
3. Conclusion
4. Maintenance of assurance

Exhibit 16 summarizes the CCEVS Preparation phase. The sponsor initiates the preparation phase by identifying the need for a PP, ST, and/or TOE evaluation to be conducted. Ideally, this determination should be made long before the PP, ST, and/or TOE is completed. The sponsor then shops around to find a CCTL with which they are comfortable doing business. Each National Evaluation Authority maintains a current list of accredited CCTLs within their jurisdiction. Annex D lists the CCTLs accredited at the time of writing by CCRA participant country. The sponsor delivers the PP, ST, and/or TOE, along with the pertinent evidence to the CCTL. After an informal review of the CC artifacts, the CCTL prepares an Evaluation Work Plan, which includes a deliverables list and an evaluation schedule. The Evaluation Work Plan identifies the activities, subactivities, work units, tasks, and subtasks to be performed by the CCTL. The deliverables list identifies all inputs needed from the sponsor in order for the CCTL to execute the Evaluation Work Plan. This list should mirror the CC evidence index supplied by the sponsor. The evaluation schedule highlights the major milestones during the evaluation and any dependencies among tasks. Next, the CCTL submits these three outputs to NIAP® for approval.

Exhibit 16. Evaluation Phases (CCEVS)—Phase 1 Preparation

Phase	Tasks	Inputs	Outputs
Preparation	• sponsor identifies need for security evaluation of IT product or system		
	• sponsor contacts CCTL to negotiate contract and initiate security evaluation		
	• sponsor provides PP, ST, and/or as built TOE to CCTL	• PP, ST, as-built TOE	
	• CCTL prepares evaluation work plan, deliverables list, and evaluation schedule	• CCTL procedures • CEM • CCEVS	• Evaluation Work Plan • Work package assessment table
	• CCTL submits required documentation to National Evaluation Authority for review		
	• National Evaluation Authority formally accepts proposed evaluation into scheme		• Validation Plan
	• Evaluation kick-off meeting		
	• Procedures and records orientation meeting (optional)		• MFR • Evaluation acceptance agreement • Approval to list evaluations in progress

Once approved, NIAP® officially accepts the evaluation into the CCEVS and generates a Validation Plan. The Validation Plan describes the role of the NIAP® validator, chief validator, and CCEVS director during the evaluation. Finally, an evaluation kickoff meeting of the CCTL, NIAP®, and sponsor is held. The agreement to accept the PP, ST, and/or TOE into the CCEVS is signed. Also signed is the approval to list the evaluation as an ongoing evaluation on NIAP®'s Web site. A memorandum for the record is written by NIAP® to document the meeting. Optionally, this meeting or another may be held to discuss the procedures by which evaluation records will be written and disseminated.

Exhibit 17 summarizes the CCEVS Conduct phase. After NIAP® gives the CCTL authorization to proceed, the CCTL conducts the evaluation while adhering to the requirements of their laboratory procedures, the CEM, and the CCEVS. If any questions arise that require clarification or any problems are discovered during the evaluation, the CCTL submits an OR to NIAP® and the sponsor (see Exhibit 14). As a parallel activity, the NIAP® validator documents the ongoing results of the evaluation in a monthly summary report, as illustrated in Exhibit 18. This report indicates any technical, management, or schedule concerns with the evaluation and reports accomplishments to date. When the evaluation is complete, the CCTL documents the results and their observations, conclusions, and recommendations in an ETR, which is sent to NIAP® and the sponsor (see Exhibit 15). Each evaluator action element that the CCTL performs results in a verdict (pass, inconclusive, or fail). A verdict is subsequently reached for each assurance component, class, and package. The overall verdict must be fail if any of the preceding constituent

Exhibit 17. Evaluation Phases (CCEVS): Phase 2 Conduct

Phase	Tasks	Inputs	Outputs
Conduct	• National Evaluation Authority gives CCTL authorization to proceed; provides technical oversight • CCTL performs evaluation of PP, ST, and/or as built TOE	• PP, ST, as built TOE • CCTL procedures • CEM • CCEVS	
	• CCTL submits ORs to sponsor and National Evaluation Authority • NIAP® validator documents results of IT security evaluation as work proceeds		• ORs • Monthly summary reports • Work package assessment table
	• CCTL completes evaluation and submits ETR to National Evaluation Authority and sponsor.		• Evaluation work package records
	• National Evaluation Authority reviews ETR to confirm validation can proceed		• ETR

Exhibit 18. Monthly Summary Report Content

1. Accomplishments
2. Outstanding Action Items
3. Technical Issues/Concerns
4. Management Issues/Concerns
5. Project Schedule
6. Project Status against Schedule
7. Validation Plan
8. Records Generated
9. Evaluation Evidence
10. Personnel
11. Improvement Suggestions
12. Validation Time

verdicts are fail, according to the CEM. Based on the overall verdict, the CCTL makes a recommendation to NIAP® as to whether or not the PP or TOE should be certified. NIAP® reviews the ETR to determine if the evaluation should continue to the next phase; this may include asking the CCTL or sponsor some questions.

Exhibit 19 summarizes the CCEVS Conclusion phase. In response to the ETR, NIAP® generates a draft Validation Report (VR), which is sent to the CCTL and sponsor for comment and review. Exhibit 20 illustrates the contents

Exhibit 19. Evaluation Phases (CCEVS): Phase 3 Conclusion

Phase	Tasks	Inputs	Outputs
Conclusion	• National Evaluation Authority reviews final ETR; addresses questions/concerns to CCTL • Sponsor and CCTL review draft VR; provide comments to National Evaluation Authority • National Evaluation Authority publishes final VR and issues CC Certificate • Validation post-mortem meeting is held (NIAP®/CCTL).	• final ETR	• draft VR • final VR • CC Certificate • EPL entry • Lessons learned report

Exhibit 20. Validation Report Content

1. Executive Summary
2. Identification
3. Security Policy
4. Assumptions and Clarification of Scope
   - 4.1 Usage
   - 4.2 Environmental
   - 4.3 Scope
5. Architectural Information
6. Documentation
7. IT Product Testing
8. Evaluated Configuration
9. Results of the Evaluation
10. Evaluator Comments/Recommendations
11. Annexes
12. Security Target
13. Glossary
14. Bibliography

of a VR. The VR is a sanitized version of the ETR which will be made public. The purpose of the VR is to: (1) confirm the CCTL findings, and (2) provide additional background information to help potential customers make an informed decision about the suitability of using this TOE in their environment. The VR is signed by the NIAP® validator, chief validator, and CCEVS director. At the same time, NIAP® issues a CC Certificate to the CCTL and the sponsor and adds an entry to the EPL. Keep in mind that CC Certificates and VRs do not have any legal status. The CCRA makes it clear that:[23]

> ...CC Certificates and VRs do not constitute or create any substantive or procedural rights, liabilities, obligations, warranty, guarantee, or endorsement by CCRA Participants.

Finally, NIAP® and the CCTL hold a postmortem meeting to discuss lessons learned about the evaluation process. Exhibits 21 and 22 illustrate the content of CC Certificates for a PP and a TOE. CC Certificates bear the logo of the National Evaluation Authority.

Exhibit 23 summarizes the CCEVS Maintenance of Assurance phase, which consists of three subphases: acceptance, monitoring, and reevaluation. Maintenance of assurance requirements (AMAs) are specified in a PP and the associated ST. The sponsor requests entry into a Certificate Maintenance Program (CMP) at the start of a formal CCTL ST evaluation. As part of this request, the sponsor submits an Assurance Maintenance Plan (AMP) and TOE component categorization report. AMA_AMP.1 defines the content of the AMP, while AMA_CAT.1 defines the content of the TOE component categorization report. The AMP describes the processes, procedures, and controls that will be followed by the developer to ensure that the EAL to which the TOE was certified is maintained during the operations and maintenance phase. In particular, flaw remediation procedures (ALC_FLR) are important. In the AMP, the developer must identify a development security analyst who has an independent reporting channel from personnel performing operations and

Exhibit 21. Content of a Common Criteria Certificate for a Protection Profile (CCEVS)

- Protection Profile Developer:
- Protection Profile Name/Identifier:
- Version Number:
- Functionality and Assurance Packages:
- Name CCTL:
- Validation Report Number:
- Date Issued:
- Signature of NIST and NSA certificate-issuing authorities:
- A statement indicating that:[102]

The Protection Profile has been evaluated at an accredited testing laboratory using the Common Methodology for Information Technology Security Evaluation (version number) for conformance to the Common Criteria for Information Technology Security Evaluation (version number).

The evaluation has been conducted in accordance with the provisions of the NIAP® Common Criteria Evaluation and Validation Scheme, and the conclusions of the testing laboratory in the evaluation technical report are consistent with the evidence presented.

The issuance of a certificate is not an endorsement of the Protection Profile by NIST, NSA, or any agency of the U.S. Government and no warranty of the Protection Profile is either expressed or implied.

The certificate applies only to the specific version of the Protection Profile as evaluated.

Exhibit 22. Content of a Common Criteria Certificate for an IT Product (CCEVS)

- Product Developer:
- Product Name:
- Type of Product:
- Version and Release Numbers:
- Protection Profile Conformance:
- Evaluation Platform:
- Name of CCTL:
- Validation Report Number:
- Date Issued;
- Assurance Level:
- Signature of NIST and NSA certificate-issuing authorities:
- A statement indicating that:[102]

The IT product has been evaluated at an accredited testing laboratory using the Common Methodology for Information Technology Security Evaluation (version number) for conformance to the Common Criteria for Information Security Evaluation (version number) as articulated in the product's functional and assurance security specification contained in its security target;

The evaluation has been conducted in accordance with the provisions of the NIAP® Common Criteria Evaluation and Validation Scheme, and the conclusions of the testing laboratory in the evaluation technical report are consistent with the evidence presented;

The issuance of a certificate is not an endorsement of the IT product by NIST, NSA, or any agency of the U.S. government and no warranty of the product is either expressed or implied;

The certificate applies only to the specific version of the product in its evaluated configuration.

maintenance functions. The TOE component categorization report ranks components relative to their importance to TOE security. The CCTL evaluates the AMP and sends its results and recommendation to NIAP®. NIAP® reviews and approves the AMP. The TOE is officially accepted into the CMP upon issuance of the initial CC Certificate.

The monitoring subphase begins immediately after the initial CC Certificate is issued. The sponsor submits proposed changes in the TOE, along with a security impact assessment of each, to NIAP®. The proposed changes may be the result of preventive, corrective, or adaptive maintenance. NIAP® verifies that the proposed changes are within the scope of the AMP and gives approval to proceed. The sponsor selects a CCTL to perform CMP-related activities. This may or may not be the same CCTL that performed the initial evaluation. The developer conducts maintenance of assurance activities according to the AMP, generating AM evidence in the process. The CCTL evaluates the evidence at regular milestones and ensures that the AMP procedures are being followed; this process includes site visits and audits. A Certificate Maintenance Summary Report (CMSR) is issued annually by the CCTL to NIAP®. A detailed Certificate Maintenance Report (CMR) is prepared by the CCTL following each major correction or enhancement. The required content of the CMR and CMSR will

Exhibit 23. Evaluation Phases (CCEVS): Phase 4 Maintenance of Assurance

Phase	Tasks	Inputs	Outputs
Maintenance of Assurance	Acceptance • sponsor requests entry into Certificate Maintenance Program at start of initial evaluation • sponsor submits AMP and related documentation to CCTL; appoints developer security analyst • CCTL evaluates AMP and related documentation as part of initial evaluation • CCTL submits EETR to National Evaluation Authority for review and approval • National Evaluation Authority reviews AMP and related documentation as part of initial evaluation TOE is officially accepted into Certificate Maintenance Program upon issuance of initial CC Certificate	• ST • TOE	• AMP • TOE Component Categorization Report
	Monitoring • sponsor submits proposed changes to National Evaluation Authority • National Evaluation Authority verifies that changes are within scope; gives approval to proceed • sponsor selects CCTL to conduct CMP related activities • developer conducts assurance maintenance activities; CCTL evaluates • CCTL conducts site visits, audits, reviews AM evidence • CCTL reports evaluation results to National Evaluation Authority • National Evaluation Authority reviews and validates CMR; issues new CC Certificate	• Proposed changes • SIA report	• AM evidence • CMR, CMSR • CC Certificate
	Re-evaluation [Return to preparation phase]		

be defined in CCEVS Publication 6. NIAP® reviews the CMR, validates it, and issues a new CC Certificate if two conditions are met:[102]

a) the security impact analysis shows that changes to the TOE are within the scope of the CMP approval, and
b) there are no outstanding non-compliances with the AMP or CMP.

If at any time during the monitoring subphase the developer or CCTL determines that the AMP must be changed, the sponsor must request a return to the acceptance subphase. If at any time during the monitoring subphase the CCTL determines that either: (1) the AMP is not being followed, or (2) the nature and extent of the proposed changes are beyond the scope of the CMP, the monitoring subphase is exited and the TOE must undergo recertification. For example, feedback from the operational environment may indicate the need to update the ST and reevaluate the TOE. A CCEVS policy letter was

Exhibit 24. Timetable for Scheduling CEM Reviews

CEM Phase	Trigger Event
Preparation	• when developer starts work on (PP evaluation) • when developer starts work on ST (ST and/or TOE evaluation)
Conduct	• when PP is complete (PP evaluation) • when ST is complete (ST or TOE evaluation)
Conclusion	N/A—schedule controlled by CCTL
Maintenance of Assurance:	• at start of ST evaluation
• Acceptance	• immediately after CC Certificate is issued
• Monitoring	• N/A, triggered by nature and extent of changes
• Re-evaluation	

issued June 7, 2002, to clarify the maximum time a product or system can stay on the in-evaluation list. In essence, the NIAP® validator will report to the CCEVS director, via a Maintenance Summary Report, any evaluations for which no activity has been reported (i.e., resolution of ORs) for two months. After the third month of inactivity the evaluation will be closed and removed from the in-evaluation list.

Exhibit 24 illustrates the optimal timetable for the sponsor to initiate the different CEM phases and the critical milestones which trigger them.

5.4 Interpretation of Results

Customers select COTS products and accept systems, including the associated documentation, based on the results of security assurance activities. Interim results from security assurance activities indicate whether or not a project is on track for meeting stated security objectives. Results from a formal CCTL evaluation of a PP, ST, or "as-built" TOE indicate whether or not the requirements of the PP and specified EAL were achieved. Results from maintenance of assurance activities indicate whether or not the assurance level to which the TOE was certified is being sustained during the in-service phase.

Customers do not participate in security assurance activities, other than for a PP. However, they are the primary consumer of the results produced therein. These results are specifically designed to:[19]

> ...help customers determine whether the IT product or system is secure enough for their intended application and whether the security risks implicit in its use are tolerable.

Evaluation results provide an objective assessment of (1) whether or not all specified SFRs and SARs have been implemented, (2) whether or not security functions have been implemented correctly, and (3) the effectiveness of these security functions to satisfy stated security objectives. In summary:[96]

> The certification/validation of evaluation results can provide a sound basis for confidence that security measures are appropriate to meet a given threat and that they are correctly implemented. However, the certification/validation of evaluation results should not be viewed as an absolute guarantee of security. Indeed, the term 'security' should always be viewed in relation to a particular set of threats and assumptions about the environment.

Vendors of COTS security products are justifiably proud of their certificates and prominently cite EAL ratings in advertisements, but what do these ratings really mean to potential customers? An EAL rating (whether for a COTS product, component TOE, or composite TOE) means the following:

- A CCTL has verified that the TOE and TOE evidence meet all SARs for the specified EAL, including augmentations and extensions.
- A CCTL has verified that the TOE is a correct, complete, and accurate implementation of SFRs stated in the ST.
- A CCTL has verified that the TOE achieves all security objectives stated in the ST.
- A CCTL has verified that the TOE adheres to all assumptions, counters all threats, and enforces all organizational security policies stated in the ST.
- All defects or deficiencies discovered during the formal evaluation have been corrected by the developer and verified by the CCTL.
- A National Evaluation Authority has validated the work performed by the CCTL.

Customers should be aware of several caveats in regard to EAL ratings. First, an EAL rating only applies to the evaluated configuration; optional features, components, and configurations are not covered unless this is explicitly stated on the CC Certificate. Second, the TOE was constructed against the developer's ST, not the customer's PP; consequently, the customer should read the ST to determine (1) which, if any, assumptions, threats, security objectives, or security requirements were added, modified, or deleted in the ST compared to those stated in the PP and why; (2) the logical and physical security boundaries; and (3) the TSF scope of control. In other words, customers are responsible for ascertaining what was and was not included within the scope of the evaluation. Third, customers need to read the final report issued by the CCTL through the National Evaluation Authority (this may be an ETR or VR, depending on what information the developer deems to be proprietary). This report clarifies the evaluated configuration, including hardware, software, firmware, and documentation; describes the evaluation methods, techniques, and tools used; details the results observed for each security assurance subactivity; provides a chronology and resolution of observation reports; and presents the conclusions and recommendations of the CCTL. A thorough reading of this report is necessary for a customer to make an informed decision about the suitability of a product or system, particularly whether or not it is appropriate for the customer's intended operational modes, scenarios, and environment.[102] As the standard notes:[19]

> A rating made relative to the CC represents the findings of a specific type of investigation of the security properties of a TOE. Such a rating does not guarantee fitness for use in any particular application environment. The decision to accept a TOE for use in a specific application environment is based on consideration of many security issues, including the CC evaluation results.

While every effort has been made to standardize evaluation results and ensure their objectiveness and repeatability, customers should be aware of some of the limitations associated with this process. First, the evaluation takes place in a laboratory, not the operational environment. This means that some issues such as latency, timing constraints, or environmental conditions may not be adequately evaluated and will need to be evaluated by another means. Second, if the item being evaluated is a COTS product or component TOE, integration issues relative to the meta-system may not be adequately evaluated and will need to be evaluated by another means. Third, the CC methodology does not include test methods explicitly related to capacity loading, saturation, or stress testing; rather, this is left to the discretion of the CCTL. Performance issues will also have to be evaluated by another means. Fourth, at present, the CC evaluation scheme only permits a pass/fail recommendation from a CCTL for a PP, ST, or "as-built" TOE. Only the results from passing vendors are released. Olthoff notes some reasons why this may not be an optimal approach:[114]

> If, however, there are no products which successfully meet a [Protection] Profile, it is in the best interests of both the user community and the vendors to allow dissemination and confirmation of the details of the evaluation results, if the vendor chooses to release them. Given a product which failed all tests and a product which failed only one test when evaluated referring to the same Protection Profile, the customer would definitely benefit from knowing which product had the better results, even if both failed. Even in cases where one product passed and one product failed by a small margin, the customer may wish to know this. A substantial price difference or the nature of the test that the one product failed may make the failed product the better buy for some applications.

5.5 Relation to Security Certification and Accreditation Activities (C&A)

Many organizations require systems and networks to undergo a security certification and accreditation process before they can transition to the operational environment. In this instance, certification refers to a:[78]

> ...comprehensive evaluation of the technical and non-technical security features of an IT system and other safeguards, made in support of the accreditation process, to establish the extent to which a particular design and implementation meets a set of specified security requirements.

A system or network is said to be accredited once a formal declaration has been made by the designated approval authority (DAA) that an IT system is approved to operate in a particular security mode using a prescribed set of safeguards to an acceptable level of risk.[78]

NSTISSP #1175 and other national directives mandate the use of CC-certified products. How do these two types of evaluations relate to each other? How can system owners avoid having to perform overlapping or duplicate security evaluations? This subsection explores these topics.

The National Information Assurance Certification and Accreditation Process (NIACAP), issued in April 2000, is the current C&A process used by the U.S. Department of Defense (DoD). NIACAP evolved from an earlier C&A process known as DoD IT Security Certification and Accreditation Process (DITSCAP), which was issued in 1997. NIACAP supports four levels of certification, reflecting different needs in regard to protecting sensitive assets and schedule and budget constraints. In contrast to the CC/CEM, the system certifier determines the appropriate evaluation level. Like the CC/CEM, NIACAP certification tasks are performed according to the certification level. The NIACAP evaluation levels and how they correspond to EALs are:

- Level 1, basic security review, is equivalent to EAL 1.
- Level 2, minimum analysis, is equivalent to EAL 2.
- Level 3, detailed analysis, is equivalent to EAL 3/4.
- Level 4, comprehensive analysis, is equivalent to EAL 4/5.

NIACAP assigns four major categories of roles and responsibilities. The program management role is responsible for the same functions as the CC/CEM sponsor and developer combined. They initiate the certification process, develop the system, correct security deficiencies, support the certification process, and maintain the C&A posture after certification. The DAA is responsible for the functions performed by the CC/CEM National Evaluation Authority, while the Certifier performs the role of the CCTL. The User Representative performs the role of the CC/CEM customer.

Like the CC/CEM, NIACAP supports an incremental verification process, the entirety of which is then subjected to validation. As shown in Exhibit 25,

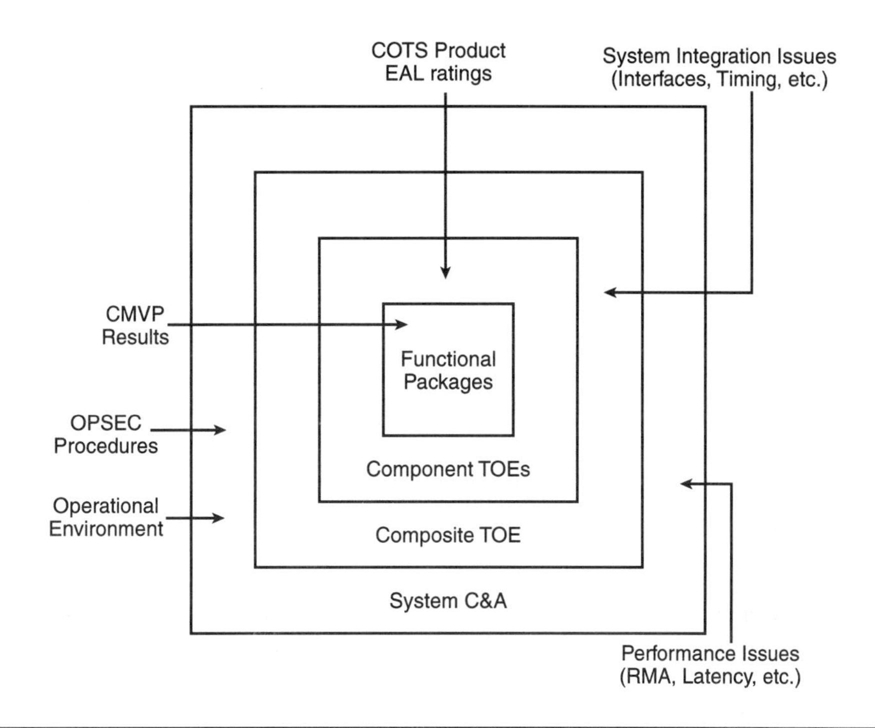

Exhibit 25. Incremental Verification Process: CC/CEM through C&A

functional packages, or modules, are verified initially. Results from other types of security testing and evaluation results, such as the Cryptographic Module Validation Program (CMVP), may be included in the assessment of functional packages. Functional packages are combined to create component TOEs, or subsystems, which are verified and given an EAL rating. A COTS product is one example. Next, component TOEs are combined to create a composite TOE that is verified. The impact of system integration issues, such as timing and interfaces, on security is evaluated. The composite TOE is assigned an EAL rating. NIACAP and other C&A processes go one step further by moving the security testing and evaluation to the operational environment. Factors not evaluated by the CC/CEM are assessed, such as the impact of RMA and latency on security, Tempest testing, contingency plans, operational procedures, and site and physical security. If the assessment is positive, the certifier makes a recommendation to the DAA to approve the system for C&A.

The National Information Assurance Certification and Accreditation Process consists of four evaluation phases, as shown in Exhibit 26. The Definition phase corresponds to the CC/CEM Preparation phase and is composed of three subphases: preparation, registration, and negotiation. The inputs to this phase parallel the information contained in a PP and ST. The two major outputs are the tailored NIACAP, which is equivalent to the CC/CEM Evaluation Work Plan, and the System Security Authorization Agreement (SSAA). The program manager generally drafts the SSAA while the certifier prepares the work plan.

Exhibit 26. Comparison between CCC/CEM and NIACAP Evaluation Phases and Artifacts

CC/CEM			NIACAP		
Phase	**Inputs**	**Outputs**	**Phase**	**Inputs**	**Outputs**
Preparation	• PP, ST, TOE	• Evaluation Work Plan • Work Package Assessment Table • Validation Plan	Definition	• Mission need • Threat Assessment • Requirements Specification • System Architecture • System Design	• System Security Authorization • NIACAP Work Plan
Conduct	• PP, ST, TOE • Security assurance evidence	• ORs • Monthly summary reports • Evaluation work package records • Work package assessment table • ETR	Verification	• SSAA • Requirements specification • System architecture • System design • CM plan	• Updated SSAA
Conclusion	• Final ETR	• Final VR • CC Certificate • EPL entry	Validation	• SSAA • Test procedures • Site information	• updated SSAA • DAA recommendation • Accreditation decision
Maintenance of Assurance	• Proposed changes • SIA report	• AM Plan • Component Categorization Report • AM evidence • CMR • CC Certificate	Post-Accreditation	• SSAA • test procedures • site information • security change impact analyses	decision whether recertification is needed

The SSAA plays a central role in all phases of a NIACAP evaluation. Initially, it documents the terms and conditions for the C&A exercise. Ultimately, it is a formal agreement among the DAA, certifier, user representative, and program manager that specifies information assurance (IA) requirements. Specifically, the SSAA defines the:[78]

- Logical and physical security boundaries
- System security architecture
- Operational environment and associated threats
- Security policies
- Accreditation requirements
- Test plans, procedures, certification results, and residual risk

The first four items correspond to the first three sections of a PP, the fifth item reflects a combination of SFRs and SARs, and the last item represents evidence generated through the performance of security assurance activities.

The second phase in a NIACAP evaluation is referred to as Verification and consists of two subphases: system development and integration, and the initial certification analysis. The NIACAP Verification phase aligns with the first half of the CC/CEM Conduct phase (ST evaluation). The SSAA is updated, and technical and lifecycle management documentation is generated by the program manager and reviewed by the certifier.

The third NIACAP evaluation phase, referred to as Validation, consists of three subphases: evaluation of the integrated system, developing a recommendation for the DAA, and making an accreditation decision. The NIACAP Validation phase aligns with the last half of the CC/CEM Conduct phase (TOE evaluation) and the CC/CEM Conclusion phase. During this phase, the additional factors not evaluated by the CC/CEM are assessed by the NIACAP. An overall risk management review is held as well.

The final NIACAP evaluation phase, Post-Accreditation, corresponds to the CC/CEM Maintenance of Assurance phase. The goal of both is to:[78]

> …certify that the information system meets documented security requirements and will continue to maintain the accredited security posture throughout the system lifecycle.

System and security operations are monitored to determine compliance with the C&A posture; in particular, security change impact analyses are validated. The certifier makes a recommendation to the DAA on a regular basis as to whether or not recertification is necessary.

In summary, all of the CC/CEM artifacts feed directly into the NIACAP; the two processes are complementary. A few additional artifacts must be generated for the NIACAP to address the factors not evaluated by the CC/CEM.

5.6 Summary

Security assurance provides confidence that a product or system will meet, has met, or is continuing to meet its stated security objectives.[19] Evidence is generated at major milestones to indicate whether or not a project is on track for achieving and sustaining security objectives. Security assurance activities are ongoing throughout the system development lifecycle, from the initiation of a PP to the certification of a TOE; verification is not a one-time event that occurs after a system or product is developed. Security assurance activities also continue during operations and maintenance to ensure that the EAL to which a system or product has been certified is maintained after initial certification and between recertification cycles.

Three sources define what security assurance activities are performed, how they are performed, and by whom. Each of these sources adds an extra level of detail, building upon the information provided by the predecessor standard. The three sources are ISO/IEC 15408-3, the CEM Parts 1 and 2, and the CC evaluation schemes promulgated by each National Evaluation Authority. ISO/IEC 15408-3 defines 10 assurance classes, 42 assurance families, and 93 assurance components. Elements that form assurance components are characterized as developer action elements, content and presentation of evidence criteria, and evaluator action elements. ISO/IEC 15408-3 also defines seven standard assurance packages: EAL 1 through 7. The standardized definitions in ISO/IEC 15408-3 permit the customer, developer, and evaluator to understand beforehand the exact criteria by which a PP, ST, and TOE will be evaluated. The CEM builds upon this foundation by specifying roles, responsibilities, activities, subactivities, actions, and work units to be fulfilled. Schemes promulgated by each National Evaluation Authority build upon the CEM by providing details about how the evaluation process is conducted and managed. National evaluation schemes may add to CEM requirements but they may not subtract from them.

The four categories of CC evaluations that are conducted are PP, ST, TOE, and maintenance of assurance, which correspond to the four generic phases of a system lifecycle or acquisition process. The key players (customer, developer, sponsor, evaluator) have different roles and responsibilities in each type of evaluation. While an ST undergoes a formal CCTL evaluation and may pass or fail, CC Certificates are not issued for STs; CC Certificates are only issued for PPs and TOEs.

The sponsor should establish a relationship with an evaluator (i.e., CCTL) prior to TOE construction. To optimize schedule performance, all parties involved (customer, sponsor, developer, and evaluator) must understand their roles and responsibilities before a formal evaluation begins. It is particularly important for developers to understand all the CC artifacts and evidence they are responsible for generating. Likewise, the customer or sponsor should ensure that their contract with the developer specifies all the necessary CC artifacts and evidence as contract deliverables. Note that the CC specifies the content of the artifacts and evidence, not the format. The customer or sponsor is responsible for defining the preferred format. Lack of this evidence and poor-quality evidence are the most common reasons for delays in completing a TOE evaluation.

Passing the initial CC certification of a product or system is a major accomplishment, but this is not the end; rather, it is the beginning of a series of ongoing security assurance activities. The operations and maintenance phase is when the real effectiveness, robustness, and resilience of the TSF are proven. Correct performance in the operational environment is the ultimate goal; this is the reason why security assurance activities are specified and undertaken in the first place. Maintenance of assurance requirements must be defined in a PP before the TOE is constructed; it is difficult to retrofit maintenance of assurance without significant cost and schedule impact. Responsibilities for performing maintenance of assurance activities (customer, developer, or third-party system integrator) need to be delineated during the development of the PP.

Customers need to be aware of several caveats in regard to EAL ratings. First, an EAL rating only applies to the evaluated configuration; optional features, components, and configurations are not covered unless this is explicitly stated on the CC Certificate. Second, the TOE was constructed against the developer's ST, not the customer's PP. Third, customers need to read the final evaluation reports.

While every effort has been made to standardize evaluation results and ensure their objectiveness and repeatability, customers should be aware of some of the limitations associated with this process. First, the evaluation takes place in a laboratory, not the operational environment. Second, if the item being evaluated is a COTS product or component TOE, integration issues relative to the meta-system may not be adequately evaluated and should be evaluated by another means. Third, the CC methodology does not include test methods explicitly related to capacity loading, saturation, or stress testing; rather, this is left to the discretion of the CCTL. Fourth, at present, the CC evaluation scheme permits only a pass/fail recommendation from a CCTL for a PP, ST, or "as-built" TOE.

Many organizations require systems and networks to undergo a security certification and accreditation process before they can transition to the operational environment. The National Information Assurance Certification and Accreditation Process, issued in 2000, is the current C&A process used by the U.S. Department of Defense. NIACAP roles, responsibilities, and evaluation phases are quite similar to the CC/CEM. In addition, all CC/CEM artifacts feed directly into the NIACAP; the two processes are complementary. A few additional artifacts must be generated for the NIACAP to address the factors not evaluated by the CC/CEM.

5.7 Discussion Problems

1. When are security assurance activities performed?
2. Who performs security assurance activities?
3. Who interprets the results of security assurance activities?

4. What is an OR, and how is it used?
5. What role do RIs play in security assurance?
6. What limitations or constraints are associated with an EAL?
7. How is a National Evaluation Authority involved in a CC evaluation?
8. Explain the similarities and differences between accredited CCTLs and CMVP laboratories.
9. How are explicit requirements verified?
10. Explain the similarities and differences between ISO/IEC 15408-3 and the CEM.
11. How are CCEVS publications used?
12. When do maintenance of assurance activities begin? When do they end?
13. Who determines what EAL a product or system is evaluated to?
14. How are CC artifacts used during certification and accreditation?
15. Several assurance components are not part of an EAL. What are they used for?
16. If no developer documentation exists, how is a TOE evaluated?
17. What is an ETR, and how is it used?
18. If no PP exists, how is a TOE evaluated?
19. Explain the similarities and differences between the CC/CEM and NIACAP.
20. When is it necessary to recertify a product or system?
21. Explain the similarities and differences between the security assurance activities performed and the roles and responsibilities involved for the initial certification of a TOE, maintenance of assurance activities, and subsequent recertifications.

Chapter 6

Postscript

The Common Criteria (CC) standard, methodology, and user community are young and as a result dynamic. The CC Implementation Management Board (CCIMB) had the foresight to put into place *a priori* a formal process to facilitate the evolution of the standard and methodology so it could stay current with the continual rapid advances in information technology. During the development of this book, three CC/CEM supplements were issued (one final, two drafts), and two new countries signed the Common Criteria Recognition Agreement (CCRA). In the United States, one CCTL was added and one CCTL was deleted from the list of accredited laboratories, and one Final Interpretation was issued. This chapter explores emerging concepts and planned events within the CC/CEM, to help the reader stay abreast of new developments. These concepts, which are under discussion within the CC user community, have not yet been formally incorporated into the standard or methodology but are likely to be so in the near future. Specifically, developments related to the following topics are discussed:

- ASE: Security Target Evaluation
- AVA: Vulnerability Analysis and Penetration Testing
- Schedules for new CC standards (ISO/IEC and CCIMB)

6.0 ASE: Security Target Evaluation

In May 2002, the Common Criteria Implementation Management Board (CCIMB) issued a draft ASE CEM supplement for public comment and review. This supplement incorporates lessons learned to date from security target evaluations. When approved, the changes proposed in the supplement will be incorporated into the CEM and Part 3 of the CC; CC Part 1 Annex B and C will be deleted, as will CC Part 3 Chapter 3. Similar changes may ripple through APE as well. In summary, the supplement proposes reorganizing and rescoping some ASE families:

- *ASE_INT* — This family will absorb the information (content and presentation of evidence) currently required by ASE_DES.
- *ASE_CCL* — This new family (conformance claims) will replace ASE_PPC. ASE_CCL will be moved up to be the second section in a Security Target and expanded to reflect CC, PP, and package conformance claims.
- *ASE_SPD* — This new family (security problem definition) will replace ASE_ENV. It contains the same three subsections: assumptions, threats, and organizational security policies.
- *ASE_OBJ* — This family will be expanded to include security objectives for the development environment.
- *ASE_SRE* — This family has been renamed extended security requirements.

6.1 AVA: Vulnerability Analysis and Penetration Testing

In July 2002, the Common Criteria Implementation Management Board (CCIMB) issued a draft AVA CEM supplement for public comment and review. This supplement incorporates lessons learned to date from the evaluation of TOE evidence. When approved, the changes proposed in the supplement will be incorporated into the CEM and Part 3 of the CC. Other minor changes may be necessary elsewhere in the CC and CEM.

In summary, the draft supplement subdivides the current AVA_VLA family into three new AVA families:

1. AVA_ALT, Attack Method Testing
2. AVA_VED, Exploitability Disposition of Potential Vulnerabilities
3. AVA_VLI, Identification of Potential Vulnerabilities.

The purpose of AVA_ALT is to determine whether or not the TSP can be undermined or exploited in the operational environment by attack methods identified in PP and ST. In fact, evaluator action item AVA_ALT.1.2E requires the evaluator to conduct penetration testing to confirm this.

The purpose of AVA_VED is to determine the exploitability of potential vulnerabilities in the operational environment that could result in a TSP violation. The three components of this family correspond to basic, medium, and high attack potentials.

The purpose of AVA_VLI is to identify potential vulnerabilities through three increasing levels of systematic searches. AVA_VLI.1 is considered a focused search that examines ADV and AGD evidence along with public information. AVA_VLI.2 is defined as a methodical search of a subset of the TOE. Both the developer and evaluator are required to perform vulnerability analyses. In contrast, AVA_VLI.3 is a complete methodical search of the entire TOE.

The new families are mapped to EALs, and it is expected that some new EALs will be defined in the future. At present, the AVA_CCA, AVA_MSU, and AVA_SOF families remain unchanged; however, the draft states that:

> ...the CCIMB is currently considering a merger of the SOF analysis and vulnerability analysis, making SOF another generic type of vulnerability to be considered.

6.2 Services Contracts

When the standard was first issued, there was some discussion about whether or not the CC/CEM could be applied to a services contract, such as a telecommunications services contract. Services contracts are becoming more common, especially in the United States, with the emphasis on outsourcing all functions except the "core business". Upon further reflection, it was determined that the services aspect of a contract is a non-issue as far as the CC are concerned. Who owns or operates the equipment is irrelevant to the CC methodology. The customer may simply elect to make the administrator guidance (AGD_ADM); user guidance (AGD_USR); and installation, generation and start-up procedures ((ADO_IGS) "available during security audits" rather than formal contractual deliverables.

6.3 Schedules for New CC Standards (ISO/IEC and CCIMB)

Several new releases of CC/CEM documents, both draft and final, are scheduled for the near-term:

- *ASE — Security Target Evaluation.* As mentioned above, a draft supplement was issued in May 2002, with a deadline for comments of August 31, 2002. The schedule for a next draft or final supplement will be determined based on the nature and extent of the comments received. Ultimately, information contained in the supplement will be incorporated into the CEM and Part 3 of the CC standard.
- *AVA — Vulnerability Analysis and Penetration Testing.* As mentioned above, a draft supplement was issued in July 2002, with a deadline for comments of October 31, 2002. The schedule for a next draft or final supplement will be determined based on the nature and extent of the comments received. Ultimately, information contained in the supplement will be incorporated into the CEM and Part 3 of the CC.
- *AMA — Maintenance of Assurance.* AMA is not currently part of the CCRA. However, some CCRA participants have developed guidance in this area as part of their national evaluation schemes. For example, in the United States, Common Criteria Evaluation and Validation Scheme (CCEVS) Publication 6 discusses a CC Certificate maintenance program.

The CCIMB is developing an AMA CEM supplement, and a draft for public comment and review is expected by the end of 2002.

- *ISO/IEC 15408, Parts 1–3.* The CCIMB is planning to issue a new version of the CC in 2003; this will include the majority of the final interpretations in effect at that time. This new CCIMB version will simultaneously be considered a committee draft to start the ISO/IEC update process. Given the time required for formal ISO/IEC comment resolution and balloting, a new version of ISO/IEC 15408, Parts 1–3, is expected during 2004.

Annex A: Glossary of Acronyms and Terms

This annex defines acronyms and terms as they are used by the Common Criteria (CC) community and in this book. Standardized definitions have been used wherever possible. When more than one standardized definition exists, multiple definitions are provided.

Acceptance phase: Start of an assurance maintenance cycle in which the developer establishes plans and procedures for assurance maintenance that are independently validated by an evaluator.[21]

Accreditation: (1) Formal recognition that a laboratory is competent to carry out specific tests or calibration or types of tests or calibrations.[110] (2) Confirmation by an accreditation body as meeting a predetermined standard of impartiality and general technical, methodological, and procedural competence.[102]

Accreditation body: Independent organization responsible for assessing the performance of other organizations against a recognized standard and for formally confirming the status of those that meet the standard.[102]

Accredited: Formal declaration by a designated approval authority that an information technology system is approved to operate in a particular security mode using a prescribed set of safeguards to an acceptable level of risk.[78]

Action: Explicitly described CC evaluator action element or one derived from a specified developer action element.[93]

Activity: Application of a CC assurance class.[93]

AISEP: Australasian Information Security Evaluation Programme.

AMP: Assurance Maintenance Plan; part of the formal assurance maintenance documentation submitted to the validation body by the sponsor of an evaluation that identifies the plans and procedures that a developer is to implement in order to ensure that the assurance that was established in the certified/validated TOE is maintained as changes are made to the target of evaluation (TOE) or its environment.[102]

Applicant: Entity (organization, individual, etc.) requesting the assignment of a register entry and entry label.[71]

Approved: Assessment by a national evaluation body as being technically competent in the specific field of IT security evaluation and formally authorized to carry out evaluations within the context of the CCEVS.[102]

Approved Laboratories List: The list of CCTLs authorized by a national evaluation authority to conduct IT security evaluations within the Common Criteria Evaluation and Validation Scheme (CCEVS).[102]

Approved test methods list: List of approved test methods maintained by a National Evaluation Authority which can be selected by a CCTL in choosing its scope of accreditation — that is, the types of IT security evaluations that it will be authorized to conduct using approved test methods.[102]

Assets: (1) Information or resources to be protected by the countermeasures of a TOE; assets may be external to the TOE but within the IT environment.[19] (2) Anything that has value to the organization.[62]

Assignment: Specification of a parameter filled in when an element is used in a Protection Profile (PP) or Security Target (ST).[19]

Assumption: Security aspects of the environment in which the TOE will or is intended to be used.[101]

Assurance: Grounds for confidence that an entity meets its security objectives.[117]

Attack potential: Perceived potential for success of an attack, should an attack be launched, expressed in terms of an attacker's expertise, resources, and motivation.[19]

Audit event: Potential security violation.[20]

Augmented: Addition of one or more assurance components from Part 3 of the CC to an EAL that is not normally part of that EAL.[19]

Authenticity: Property that ensures that the identity of a subject or resource is the one claimed. Authenticity applies to entities such as users, processes, systems, and information.[62]

Authorized user: User who may, in accordance with the TOE security policy (TSP), perform an operation.[19]

Availability: (1) Property of being accessible and usable upon demand by an authorized entity.[62] (2) Prevention of unauthorized withholding of information resources.[104]

Baseline controls: A minimum set of safeguards established for a system or organization.[62]

BSI: (1) *Bundesamt für Sicherheit in der Informationstechnik* (German Information Security Agency). (2) British Standards Institute.

C&A: (1) Certification and accreditation. (2) Certification and authorization to operate (FAA).

CB: Certification/validation body, an organization responsible for carrying out certification/validation and for overseeing the day-to-day operation of an evaluation and certification or validation scheme.[23]

CC: Common Criteria, common name used for the methodology defined in ISO/IEC 15408 Parts 1 to 3.

CCEB: CC Editing Board; developed first draft of CC.

CCEL: Common Criteria Evaluation Laboratory.

CCEVS: Common Criteria Evaluation and Validation Scheme. (1) Administrative and regulatory framework under which the CC is applied by an evaluation authority within a specific community.[19] (2) Systematic organization of the functions of evaluation and certification/validation under the authority of a CB in order to ensure that high standards of competence and impartiality are maintained and that consistency is achieved.[23] (3) Set of rules defining the evaluation environment, including criteria and methodology required to conduct IT security evaluations.[24] (4) Program developed by the National Information Assurance Partnership (NIAP®) establishing an organizational and technical framework to evaluate the trustworthiness of IT products and protection profiles.[102]

CCIMB: (1) CC Implementation Management Board, which conducted trial evaluations of first draft of CC and developed second draft of CC. (2) CC Interpretation Management Board, which renders CC interpretations to facilitate consistent evaluation results under the Common Criteria Recognition Agreement (CCRA).

CCRA: Common Criteria Recognition Agreement; *see also* MRA.

CCTL: Common Criteria Testing Laboratory, an IT security evaluation facility accredited and approved by a National Evaluation Authority to conduct CC-based evaluations.

CEM: Common Evaluation Methodology; *see* CM.

CEMEB: Common Evaluation Methodology Editing Board, CCRA participants involved in development of CEM.

Certificate authorizing participant: National Evaluation Authority and CCRA signatory that issues CC Certificates and recognizes those issued by other National Evaluation Authorities.

Certificate consuming participant: National Evaluation Authority and CCRA signatory that recognizes CC Certificates issued by other National Evaluation Authorities but at present does not issue any certificates itself.

Certificate of Accreditation: Document issued by the National Voluntary Laboratory Accreditation Program (NVLAP®) or other national evaluation authority to a laboratory that has met the criteria and conditions for accreditation. A current Certificate of Accreditation may be used as proof of accredited status and is always accompanied by a Scope of Accreditation.[110]

Certification/validation: (1) Process carried out by a CB leading to the issuance of a CC Certificate;[23] (2) comprehensive evaluation of the technical and nontechnical security features of an IT system and other safeguards, made in support of the accreditation process, to establish the extent to which a particular design and implementation meets a set of specified security requirements.[78]

Certification/validation report: Public document issued by a CB that summarizes the results of an evaluation and confirms the overall results — that is, the evaluation has been properly carried out; the evaluation criteria, evaluation methods, and other procedures have been correctly applied; and the conclusions of the Evaluation Technical Report are consistent with evidence adduced.[23,102]

Certified TOE: (1) Product or system and its associated guidance that, having been a TOE under evaluation, has completed the evaluation, its ST, certification report, and certificate having been published.[26] (2) Version of TOE that was evaluated, awarded a CC Certificate, and is listed in an evaluation authority's Evaluated Products List.[21]

Certified/Validated Products List: Public document that summarizes and confirms the results of an evaluation and lists current valid CC Certificates in accordance with the CCRA.[23,102]

CESG: Communications–Electronics Security Group (U.K.).

Check: Similar to, but less rigorous than, confirm or verify; a quick determination to be made by the evaluator, perhaps requiring only a cursory analysis, or perhaps no analysis at all.[21]

Class: Grouping of security requirements that share a common focus; members of a class are termed families.[19]

CLEF: Common Criteria Licensed Evaluation Facility.

CM: (1) Common Methodology for Information Technology Security Evaluation, a technical document that describes a particular set of IT security evaluation methods, also referred to as CEM.[23,102] (2) Configuration management, as in ACM.

CMP: Certificate Maintenance Program; a program within the CCEVS that allows a sponsor to maintain a CC Certificate by providing a means to ensure that a validated TOE will continue to meet its Security Target as changes are made to the IT product or its environment.[102]

CMR: Certificate Maintenance Report, a report prepared by a CCTL for the evaluation authority detailing the results of their evaluation maintenance activities conducted on behalf of a sponsor.[102]

CMSR: Certificate Maintenance Summary Report, an annual report prepared by a sponsor for the evaluation authority providing a summary of all certificate maintenance activities conducted during the previous year.[102]

CMT LAP: NVLAP® cryptographic module testing laboratory accreditation program.[111]

CMV: Cryptographic module validation; the act of determining if a cryptographic module conforms to the requirements of FIPS PUB 140-2.[111]

CMVP: Cryptographic Module Validation Program, a program run jointly by the Communication Security Establishment (CSE) and National Institutes of Standards and Technology (NIST) that focuses on security conformance testing of a cryptographic module against FIPS PUB 140-2, Security Requirements for Cryptographic Modules, and other related cryptographic standards.[83]

Coherent: Entity that is logically ordered and has a discernible meaning; for documentation, this adjective addresses both the actual text and the structure of the document, in terms of whether it is understandable by its target audience.[21]

Common Criteria Certificate: (1) Public document issued by a compliant CB and authorized by a participant that confirms that a specific IT product or Protection Profile has successfully completed evaluation by an IT security evaluation facility (ITSEF); a CC Certificate always has associated with it a certification and validation report.[23] (2) Formal recognition by the NIAP® validation body that the IT security evaluation has been conducted in accordance with the CCEVS requirements using the CC and CM. A product that has received a CC Certificate is placed on NIAP®'s Validated Products List.[111]

Complete: All necessary parts of an entity have been provided. In terms of documentation, this means that all relevant information is covered in the documentation, at such a level of detail that no further explanation is required at that level of abstraction.[21]

Components: Specific set of security requirements that are constructed from elements; the smallest selectable set of elements that may be included in a PP, an ST, or a package.[19]

Component TOE: TOE that forms part of a composite TOE; the lowest level TOE in an IT product or system.

Composite TOE: TOE composed of multiple component TOEs; the highest level TOE in an IT product or system.

COMPUSEC: Computer security; preventing, detecting, and minimizing the consequences of unauthorized actions by users (authorized and unauthorized) of a computer system.[99]

COMSEC: Communications security; measures and controls taken to deny unauthorized persons information derived from telecommunications and to ensure the authenticity of such telecommunications. COMSEC includes cryptosecurity, transmission security, emissions security, and physical security of COMSEC material.[78]

Confidentiality: Property that information is not made available or disclosed to unauthorized individuals, entities, or processes.[62]

Confirm: To review in detail in order to make an independent determination of sufficiency, with the level of rigor required depending on the nature of the subject matter; applicable to evaluator actions.[21]

Connectivity: Property of the TOE that allows interaction with IT entities external to the TOE. This includes exchange of data by wire or by wireless means, over any distance in any environment or configuration.[19]

Consistent: Relationship between two or more entities, indicating that there are no apparent contradictions between these entities.[21]

Corrective security objective: Security objectives that require the TOE to take action in response to potential security violations or other undesirable events, in order to preserve or return to a secure state and/or limit any damage caused.[22]

COTS: Commercial "off the shelf".

Counter: Offset, nullify, defensive response (i.e., a security objective that mitigates a particular threat but does not necessarily indicate that the threat is completely eradicated as a result).[21,30]

Cryptographic algorithm testing: Input/output testing to determine whether the implementation conforms to the specification.[111]

Cryptographic boundary: Explicitly defined contiguous perimeter that establishes the physical bounds of a cryptographic module.[111]

Cryptographic module: Set of hardware, software, firmware, or a combination thereof that implements cryptographic logic or processes, including cryptographic algorithms and key generation, and is contained within the cryptographic boundary of the module.[111]

CSE: Communications Security Establishment (Canada).

CSTT: Cryptographic support test tool; used as part of Cryptographic Module Validation Program (CMVP).

CTCPEC: Canadian Trusted Computer Product Evaluation Criteria.

Current version of TOE: Version of TOE that differs in some respect from the certified version, such as (1) a new release of the TOE, (2) a certified version with patches to correct subsequently discovered bugs, and (3) the same basic version of the TOE but on a different hardware or software platform.[21]

Data integrity: Property that data has not been altered or destroyed in an unauthorized manner.[62]

Demonstrate: Analysis leading to a conclusion; less rigorous than a proof.[21]

Dependency: Relationship between requirements such that the requirement that is depended upon must normally be satisfied for the other requirements to be able to meet their objectives.[19]

Depth: Level of design and implementation that is being evaluated.[21]

Describe: Provide specific details about an entity.[21]

Detective security objective: Security objectives that provide the means to detect and monitor the occurrence of events relevant to the secure operation of the TOE.[22]

Determine: Conducting an independent analysis, usually in the absence of any previous analysis having been performed, with the objective of reaching a particular conclusion; differs from confirm or verify, as these terms imply that an analysis has already been performed that must be reviewed.[21]

EAL: Evaluation assurance level, a package consisting of assurance components from Part 3 of the CC that represents a point on the CC predefined assurance scale. At present, the CC defines seven hierarchical EALs, from EAL 1 to EAL 7; the higher EALs encompass the requirements of the lower EALs.[19,24,112]

EAP: Evaluation Acceptance Package.

ECMA: European Computer Manufacturers' Association.

EF: Evaluation facility, an organization that carries out evaluations independently of the developers of the IT products or protection profiles, usually on a commercial basis.[23]

Element: Indivisible security requirement that can be verified by the evaluation; lowest level security requirement from which components are constructed.[19]

EMSEC: Emissions security, protection resulting from measures taken to deny unauthorized persons information derived from the interception and analysis of compromising emanations from crypto equipment or IT systems.[78]

Entry label: Naming information that uniquely identifies a registered PP or package.[71]

ESR: Evaluation Summary Report, a report issued by an overseer and submitted to an evaluation authority that documents the oversight verdict and its justification.[24]

ETR: Evaluation Technical Report; (1) a report giving details of the findings of an evaluation, submitted by the evaluation facility to the certification/validation body as the principal basis for the certification/validation report.;[23] (2) A report produced by the evaluator and submitted to an overseer that documents the overall verdict and its justification.[24]

Evaluation: (1) Assessment of a PP, ST, or a TOE against defined criteria.[19] (2) Assessment of an IT product or a protection profile against CC requirements using CEM to determine whether or not the claims made are justified.[23,112] (3) Assessment of an IT product against the CC using the CEM to determine whether or not the claims made are justified or the assessment of a PP against the CC using the CEM to determine if the profile is complete, consistent, technically sound, and hence suitable for use as a statement of requirements for one or more TOEs that may be evaluated.[104]

Evaluation authority: National body that implements the CC for a specific community by means of an evaluation scheme and thereby sets the standards and monitors the quality of evaluations conducted by CBs within that community.[19]

Evaluation scheme: *See* CCEVS.

Evaluation work plan: Document produced by a CCTL detailing the organization, schedule, and planned activities for an IT security evaluation.[102]

Evaluator action element: Assurance requirement stated in Part 3 of the CC that represents a TOE evaluator's responsibilities in verifying the security claims made in the Security Target of a TOE.[24]

Exhaustive: Used to describe the conduct of an analysis or other activity; related to systematic but considerably stronger in that it indicates not only that a methodical approach has been taken to perform the analysis or activity according to an unambiguous plan but also that the plan followed is sufficient to ensure that all possible avenues have been exercised.[21]

Explicit requirements: Functional security requirements or security assurance requirements specified in a PP or ST that satisfy a specific consumer need but do not originate from the CC catalog of standardized components (*see also* Refinement and Extended).

Extended: Addition to an ST or PP of requirements not contained in Part 2 or assurance requirements not contained in Part 3 of the CC;[19] extensibility (*see also* Explicit requirements and Refinement).

External IT entity: Any IT product or system, untrusted or trusted, outside of the TOE that interacts with the TOE.[19]

Family: Grouping of security requirements that share security objectives but may differ in emphasis or rigor; the members of a family are termed components.[19]

FIPS: Federal information processing standards; issued by NIST.

Formal: Expressed in a restricted syntax language with defined semantics based on well-established mathematical concepts.[117]

GMITS: Guidelines for the Management of IT Security.

Hierarchy: Ordering of components within a family to represent increasing strength or capability of security requirements that share a common purpose; on occasion, partial ordering is used to illustrate the relationship between nonhierarchical sets.[19]

IA: Information assurance.

IATF: Information assurance technical framework.

IATFF: Information Assurance Technical Framework Forum.

IEC: International Electrotechnical Commission.

Informal: Expressed in a natural language.[117]

INFOSEC: Information security.

Input task: Tasks related to the management of all required, sponsor-supplied evaluation evidence.[93]

Integrity: Prevention of unauthorized modification of information.[104]

Internal communication channel: Communication channel among different parts of a TOE.

Interpretation: Expert technical judgment, when required, regarding the meaning or method of application of any technical aspect of the criteria or the methodology.[23,102]

Inter–TSF transfers: Communicating data between the TOE and the security functions of other trusted IT products.[19]

ISO: International Organisation for Standardisation.

Iteration: Use of an element more than once with varying parameters.[19]

IT product: Package of IT hardware, software, and firmware that provides functionality designed for use or incorporation within a multiplicity of systems. An IT product can be a single product or multiple IT products configured as an IT system, network, or solution to meet specific customer needs. In either case, the testing occurs in a testing facility or a client's site under laboratory conditions, and not in the actual operational environment.[110]

ITSEC: Information technology security evaluation criteria; a compilation of the information that must be provided and of the actions that must be taken in order to give grounds for the confidence that evaluations will be carried out effectively and to a consistent standard throughout an evaluation and certification/validation scheme.[23,102]

IT security: All aspects related to defining, achieving, and maintaining confidentiality, integrity, availability, accountability, authenticity, and reliability.[62]

IT security policy: Rules, directives, and practices that govern how assets, including sensitive information, are managed, protected, and distributed within an organization and its IT systems.[62]

ITSEF: IT security evaluation facility, an accredited EF, licensed or approved to perform evaluations within the context of a particular IT security evaluation and certification/validation scheme.[23]

ITSEM: *Information Technology Security Evaluation Manual*, used with ITSEC.

IT security evaluation methods: Compilation of the methods that need to be used by EFs in applying ITSEC in order to give grounds for confidence that evaluations will be carried out effectively and to a consistent standard throughout an evaluation and certification/validation scheme.[23]

Justification: Analysis leading to a conclusion but which is more rigorous than a demonstration; requires significant rigor in terms of very carefully and thoroughly explaining every step of a logical argument.[21]

Monitoring of evaluations: Procedure by which representatives of a CB observe in progress or review completed evaluations in order to satisfy themselves that an ITSEF is carrying out its functions in a proper and professional manner.[23]

Monitoring phase: Middle of an assurance maintenance cycle during which the developer provides evidence at one or more points that assurance of the TOE is being maintained in accordance with established plans and procedures; this evidence is independently validated by an evaluator.[21]

MRA: CC Project Mutual Recognition Arrangement; *see* CCRA.

NATA: National Australian Testing Authority.

NIACAP: National Information Assurance Certification and Accreditation Process.

NIAP®: National Information Assurance Partnership (U.S.).

NIST: National Institute of Standards and Technology (U.S.).

NSA: National Security Agency (U.S.).

NSI: National Security Information, information that has been determined, pursuant to Executive Order 12958 or any predecessor order, to require protection against unauthorized disclosure.[78]

NSTISSAM: National Security Telecommunications and Information Systems Security Advisory/Information Memorandum.

NSTISSC: National Security Telecommunications and Information Systems Security Committee (U.S.).

NSTISSI: National Security Telecommunications and Information Systems Security Instruction.

NSTISSP: National Security Telecommunications and Information Systems Security Policy.

NVLAP®: National Voluntary Laboratory Accreditation Program; U.S. accreditation authority for CCTLs operating within the NIAP® CCEVS.

Object: Passive entity within the TOE security function (TSF) scope of control (TSC) that contains or receives information and upon which subjects perform operations.[19]

OD: Observation Decision.

ODRB: Observation Decision Review Board.

OECD: Organization for Economic Cooperation and Development.

OPSEC: Operations security; the implementation of standardized operational security procedures that define the nature and frequency of the interaction between users, systems, and system resources, the purpose of which is to (1) maintain a system in a known secure state at all times, and (2) prevent accidental or intentional theft, destruction, alteration, or sabotage of system resources.[99]

OR: Observation Report, written by the evaluator requesting a clarification or identifying a problem during the conduct of an IT security evaluation.[24,102]

OSP: Organizational security policy; one or more security rules, procedures, practices, or guidelines imposed by an organization upon its operations.[19]

Output task: Tasks related to the reporting of information through either an Observation Report or Evaluation Technical Report.[93]

Package: Set of either functional or assurance components (e.g., an EAL), combined together to satisfy a subset of identified security objectives; packages are intended to be used to build PPs and STs.[19]

PP: Protection Profile. (1) Formal document defined in the CC that expresses an implementation-independent set of security requirements for an IT product that meets specific consumer needs.[19,23,112] (2) Complete combinations of security objectives and functional and assurance requirements with associated rationale.[24]

Preventive security objective: Security objectives that prevent a threat from being carried out or limit the ways in which it can be carried out.[22]

Principal SAR: Security assurance requirement that directly contributes to assuring that an entity meets its security objectives.

Principal SFR: Security functional requirement that directly satisfies the identified security objectives of the TOE.[22]

Profile: Structure that characterizes the behavior of users and subjects; it represents how users and subjects interact with the TSF.[20]

Profile metrics: Ways in which various types of user and subject activities are recorded and measured in a profile; serves as input to pattern recognition.[20]

Profile target group: One or more users who interact with the TSF, supposedly according to historical patterns or patterns of expected behavior.[20]

Protected information: Information gathered or obtained during an evaluation, the unauthorized disclosure of which could reasonably be expected to cause: (1) harm to competitive commercial or proprietary interests, (2) a clearly unwarranted invasion of personal privacy, (3) damage to national security, or (4) harm to an interest protected by national law, legislation, regulation, policy, or official obligation.

Prove: Formal analysis in the mathematical sense which is completely rigorous in all ways.[21]

RA: Registration Authority.[71]

Recognition of CC Certificates: Acknowledgment that the evaluation and certification processes carried out by compliant CBs appear to have been carried out in a duly professional manner and meet all the conditions of the CCRA and the intention to give all resulting CC Certificates equal weight.[23]

Reevaluation: Evaluation of a new version of the TOE that addresses all security-relevant changes made to the certified version of the TOE and reuses previous evaluation results where they are still valid.[21]

Reevaluation phase: Completion of the assurance maintenance cycle in which an updated version of the TOE is submitted for reevaluation based on changes affecting the TOE since the certified version.[21]

Reference monitor: Concept of an abstract machine that enforces TOE access control policies.[19]

Reference validation mechanism: Implementation of the reference monitor concept that possesses the following properties: tamper-proof, always invoked, and simple enough to be subjected to thorough analysis and testing.[19]

Refinement: Addition of extra details to an element when it is used in a PP or ST (*see also* Explicit requirement and Extended).[19]

Reliability: Property of consistent intended behavior and results.[62]

Residual risk: (1) Portion of risks remaining after security measures have been applied.[78] (2) Risk that remains after safeguards have been implemented.[62]

Revocation: Removal of the accredited status of a laboratory if the laboratory is found to have violated the terms of its accreditation.[110]

RI: Request for Interpretation, submitted by evaluation authorities to CCIMB; the four types are (1) perceived error such that some content in the CC or CEM requires correction, (2) identified need for some additional material in the CC or CEM, (3) proposed method for applying the CC or CEM in a specific circumstance for which endorsement is sought, and (4) request for information to assist with understanding the CC or CEM.[117]

Rigor: Degree of structure and formality applied to the evaluation by the evaluators.[21]

Risk: (1) Combination of the likelihood that a threat will be carried out and the severity of the consequences should it happen.[99] (2) Potential that a given threat will exploit vulnerabilities of an asset or group of assets to cause loss or damage to the assets.[62]

Risk assessment: (1) Process of analyzing threats to and vulnerabilities of an IT system and the potential impact the loss of information or capabilities of a system would have; the resulting analysis is used as the basis for identifying appropriate and cost-effective countermeasures.[78] (2) Process of identifying security risks, determining their magnitude, and identifying areas requiring safeguards.[62]

Risk management: (1) Process concerned with the identification, measurement, control, and minimization of security risks in IT systems to a level commensurate with the value of the assets protected.[78] (2) The entire process of identifying, controlling, and eliminating or minimizing uncertain events that may affect IT system resources.[62]

Role: Predefined set of rules establishing the allowed interactions between a user and the TOE.[19]

Safeguard: Practice, procedure, or mechanism that reduces risk.[62]

SAR: Security assurance requirement.

Scope: Portion of an IT product or system that is being evaluated.[21]

Scope of Accreditation: Approved test methods for which a CCTL has been accredited.[110]

Security assurance: Grounds for confidence that an entity meets its security objectives.[19]

Security attribute: Information associated with users, subjects, and objects used for the enforcement of the TSP.[20]

Security classification: Labeling applied to protected information to indicate minimum standards of protection that need to be applied in the national or organizational interest; also referred to as protective marking.

Security flaw: Condition that alone or in concert with others provides an exploitable vulnerability. TSP violations that occur not from a problem with the hardware, software, or firmware portion of a TOE but from a problem in the TOE guidance are also recognized as security flaws.[21]

Security objective: Statement of intent to counter identified threats and/or satisfy identified organization policies and assumptions.[19]

SEI: Software Engineering Institute.

Selection: Specification of one or more items that are to be selected from a list given in the element definition.[20]

Semiformal: Expressed in a restricted syntax language with defined semantics.[117]

Sensitive information: Any information for which the loss, misuse, or unauthorized access to or modification of could adversely affect the national interest or conduct of federal programs or the privacy to which individuals are entitled under the Privacy Act Section 552a of Title 5 USC, but which has not been specifically authorized under criteria established by an Executive Order or an Act of Congress to be kept secret in the interest of national defense or foreign policy.[84]

SF: Security function; part or parts of the TOE that have to be relied upon for enforcing a closely related subset of the rules from the TSP.[117]

SFP: Security function policy; the security policy enforced by an SF.[20]

SFR: Security functional requirement.

SOF: Strength of function; a qualification of a TOE security function expressing the minimum efforts assumed necessary to defeat its expected security behavior by directly attacking its underlying security mechanisms.[19]

SOF-basic: Level of the TOE strength of function where analysis shows that the function provides adequate protection against causal breach of the TOE security by attackers possessing a low attack potential.[19]

SOF-high: Level of the TOE strength of function where analysis shows that the function provides adequate protection against deliberately planned or organized breach of the TOE security by attackers possessing a high attack potential.[19]

SOF-medium: Level of the TOE strength of function where analysis shows that the function provides adequate protection against straightforward or intentional breach of the TOE security by attackers possessing a moderate attack potential.[19]

SSAA: System Security Authorization Agreement.[78]

SSE–CMM: System Security Engineering Capability Maturity Model.

ST: Security Target; complete combination of security objectives, functional and assurance requirements, summary specifications and rationale to be used as the basis for evaluation of an identified TOE.[19,24,112]

Subactivity: Application of a CC assurance component.[93]

Subject: Active entity within the TSC that causes operations to be performed.[19]

Subtask: Subdivision of a task.[24]

Supporting SAR: Security assurance requirement that indirectly contributes to assuring that an entity meets its security objectives.[22]

Supporting SFR: Security functional requirement that does not directly satisfy security objectives for the TOE but which provides support to the principal SFRs and hence indirectly helps satisfy TOE security objectives.[22]

System integrity: Property that a system performs its intended function in an unimpaired manner, free from deliberate or accidental unauthorized manipulation of the system.[62]

Task: Specifically required CEM evaluation work that is not derived directly from a CC requirement.[24]

TCSEC: Trusted Computer System Evaluation Criteria.

Test method: Evaluation assurance package from the CC and the associated evaluation methodology from that assurance package from the CM.[110]

Threat: (1) Any circumstance or event with the potential to harm an IT system through unauthorized access, destruction, disclosure, modification of data, and/or denial of service.[78] (2) Potential danger that a vulnerability may be exploited intentionally, triggered accidentally, or otherwise exercised.[109] (3) A potential cause of an unwanted incident which may result in harm to a system or organization.[62]

TOE: Target of evaluation; an IT product or system and its associated administrator and user guidance documentation that is the subject of an evaluation.[19,23,24,112]

TOE guidance: Administrator guidance, user guidance, flaw remediation guidance, delivery procedures, and installation, generation, and start-up procedures.[26]

TOE user: Focal point in the user organization that is responsible for receiving and implementing fixes to security flaws. This is not necessarily an individual user but may be an organizational representative who is responsible for the handling of security flaws.[26]

Trusted channel: Means by which a TSF and a remote trusted IT product can communicate with necessary confidence to support the TSP.[19]

Trusted path: Means by which a user and a TSF can communicate with necessary confidence to support the TSP.[19]

TSC: TSF scope of control; the set of interactions that can occur with or within a TOE and are subject to the rules of the TSP.[20,117]

TSF: TOE security functions; a set consisting of all hardware, software, and firmware of the TOE that must be relied upon for the correct enforcement of the TSP.[19]

TSFI: TSF interface; the set of interfaces, whether interactive man–machine interfaces or application program interfaces, through which resources are accessed that are mediated by the TSF or information obtained from the TSF.[20]

TSP: TOE security policy; a set of rules that regulate how assets are managed, protected, and distributed within a TOE.[19]

TSS: TOE summary specification.

TTP: Trusted third party.

UKAS: U.K. Accreditation Society.

Users: ISO/IEC recognizes two types of authorized users: (1) local or remote human users, and (2) external IT entities. Users are considered to be outside a TOE and interact with a TOE through the TSFI.[20]

Validation: (1) Review of an IT security evaluation by an evaluation authority to determine if issuance of a CC Certificate is warranted.[111] (2) Process of applying specialized security test and evaluation procedures, tools, and equipment needed to establish acceptance for joint usage of an IT system by one or more departments or agencies and their contractors.[78]

Verification: (1) Confirmation by examination and provision of objective evidence that specified requirements have been fulfilled.[110] (2) Process of comparing two levels of an IT system specification for proper correspondence, such as security policy model with top-level specification, top-level specification with source code, source code with object code.[78]

Verify: Independent evaluator actions; similar to *confirm* but more rigorous.[21]

VID: Validation identification number.

VR: Validation Report, a publicly available document issued by a National Evaluation Authority that summarizes the results of an evaluation and confirms the overall results.[104]

Vulnerability: Weakness in the design, operation, or operational environment of an IT system or product that can be exploited to violate the intended behavior of the system relative to safety, security, and/or integrity.[62,99,109]

Work units: Smallest unit of an evaluation action; derived from an evaluator action element or a content and presentation of evidence element.[24]

Annex B: Additional Resources

This collection of additional resources lists the sources that were used during the development of this book and provides pointers to additional resources that may be of interest to the reader. It is organized in three parts: (1) standards, regulations, and policy; (2) publications; and (3) online resources.

Standards, Regulations, and Policy

This section lists historical and contemporary standards related to various aspects of the Common Criteria and their development and use. Given that most national and international standards are reaffirmed, updated, or withdrawn on a three- to five-year cycle, for implementation or assessment purposes, one should always verify that one has the current approved version.

Historical

1. DoD 5200.28-M, *ADP Computer Security Manual: Techniques and Procedures for Implementing, Deactivating, Testing, and Evaluating Secure Resource-Sharing ADP Systems*, U.S. Department of Defense, January 1973.
2. DoD 5200.28-M, *ADP Computer Security Manual: Techniques and Procedures for Implementing, Deactivating, Testing, and Evaluating Secure Resource-Sharing ADP Systems, with 1st Amendment*, U.S. Department of Defense, June 25, 1979.
3. CSC-STD-001-83, *Trusted Computer System Evaluation Criteria (TCSEC)*, National Computer Security Center, U.S. Department of Defense, August 15, 1983.
4. DoD 5200.28-STD, *Trusted Computer System Evaluation Criteria (TCSEC)*, National Computer Security Center, U.S. Department of Defense, December 1985.
5. CSC-STD-003-85, *Guidelines for Applying the Trusted Computer System Evaluation Criteria (TCSEC) in Specific Environments*, National Computer Security Center, U.S. Department of Defense, June 1985.

6. CSC-STD-004-85, *Technical Rationale Behind CSC-STD-003-83*, National Computer Security Center, U.S. Department of Defense, 1985.
7. NCSC-TG-025, version 2, *A Guide to Understanding Data Remembrance in Automated Information Systems (AIS)*, National Computer Security Center, U.S. Department of Defense, September 1991.
8. NCSC-TG-005, version 1, *Trusted Network Interpretation of the TCSEC*, National Computer Security Center, U.S. Department of Defense, July 1987.
9. NCSC-TG-011, version 1, *Trusted Network Interpretation of the TCSEC*, National Computer Security Center, U.S. Department of Defense, August 1, 1990.
10. NCSC-TG-021, version 1, *Trusted DBMS Interpretation of the TCSEC*, National Computer Security Center, U.S. Department of Defense, April 1991.
11. DDS-2600–6243–91, version 1, *Compartmented-Mode Workstation Evaluation Criteria*, Defense Intelligence Agency, U.S. Department of Defense, 1991.
12. Federal Criteria for Information Technology Security, version 1.0 (Vols. I and II), jointly published by the U.S. National Institute of Standards and Technology and National Security Agency, December 1992.
13. *Information Technology Security Evaluation Criteria (ITSEC)*, version 1.2, Office for Official Publications of the European Communities, June 1991.
14. *Information Technology Security Evaluation Manual (ITSEM)*, Office for Official Publications of the European Communities, 1992.
15. *Secure Information Processing versus the Concept of Product Evaluation*, Technical Report ECMA TR/64, European Computer Manufacturers' Association, December 1993.
16. *Guidelines for the Security of Information Systems*, Organization for Economic Cooperation and Development (OECD), November 1992.
17. *The Canadian Trusted Computer Product Evaluation Criteria (CTCPEC)*, Canadian System Security Centre, Communications Security Establishment, version 3.oe, 1993.
18. *UKSP01, UK IT Security Evaluation Scheme: Description of the Scheme*, Communications-Electronics Security Group, March 1991.

Current

19. ISO/IEC 15408-1(1999-12-01), Information Technology — Security Techniques — Evaluation Criteria for IT Security — Part 1: Introduction and General Model.
20. ISO/IEC 15408-2(1999-12-01), Information Technology — Security Techniques — Evaluation Criteria for IT Security — Part 2: Security Functional Requirements.
21. ISO/IEC 15408-3(1999-12-01), Information Technology — Security Techniques — Evaluation Criteria for IT Security — Part 3: Security assurance requirements
22. ISO/IEC PDTR 15446(2001-04), Information Technology — Security Techniques — Guide for the Production of Protection Profiles and Security Targets.
23. *Arrangement on the Recognition of Common Criteria Certificates in the Field of Information Technology Security*, May 23, 2000.
24. CEM-97/017, *Common Methodology for Information Technology Security Evaluation, Part 1: Introduction and General Model*, version 0.6, November 1997.
25. CEM-99/045, *Common Methodology for Information Technology Security Evaluation, Part 2: Evaluation Methodology*, version 1.0, August 1999.
26. CC/CEM supplements: (a) CEM-2001/0015R, *Common Methodology for Information Technology Security Evaluation, Part 2: Evaluation Methodology, Supplement: ALC_FLR — Flaw Remediation*, version 1.1, February 2002; and (b) CCIMB-2002-04-011, *Common Methodology for Information Technology Security Evaluation, Supplement: ASE — Security Target Evaluation*, (draft) version 0.6, May 2002.

27. CCIMB Final Interpretation-004, ACM_SCP.*.1C Requirements Unclear, November 12, 2001.
28. CCIMB Final Interpretation-006, Virtual Machine Description, October 15, 2000.
29. CCIMB Final Interpretation-Augmented and Conformant Overlap, July 31, 2001.
30. CCIMB Final Interpretation-009, Definition of "Counter," April 13, 2001.
31. CCIMB Final Interpretation-013, Multiple SOF Claims for Multiple Domains in a Single TOE, October 15, 2000.
32. CCIMB Final Interpretation-024, Required Evaluation Evidence for Commercial "Off the Shelf" (COTS) Products, February 16, 2001.
33. CCIMB Final Interpretation-025, Level of Detail Required for Hardware Descriptions, July 31, 2001.
34. CCIMB Final Interpretation-027, Events and Functions in AGD_ADM, February 16, 2001.
35. CCIMB Final Interpretation-031, Obvious Vulnerabilities, February 16, 2001.
36. CCIMB Final Interpretation-032, Strength of Function Analysis in ASE_TSS, October 15, 2000.
37. CCIMB Final Interpretation-033, Use of "Check" in Part 3, October 15, 2000.
38. CCIMB Final Interpretation-037, ACM on Product or TOE?, February 16, 2001.
39. CCIMB Final Interpretation-043, Meaning of "Clearly Stated" in APE/ASE_OBJ.1, February 16, 2001.
40. CCIMB Final Interpretation-049, Threats Met by Environment, February 16, 2001.
41. CCIMB Final Interpretation-055, Incorrect Components Referenced in Part 2 Annexes, FPT_RCV, October 15, 2000.
42. CCIMB Final Interpretation-058, Confusion Over Refinement, July 31, 2001.
43. CCIMB Final Interpretation-062, Confusion Over Source of Flaw Reports, July 13, 2001.
44. CCIMB Final Interpretation-064, Apparent Higher Standard for Explicitly Stated Requirements, February 16, 2001.
45. CCIMB Final Interpretation-065, No Component to call out Security Function Management, July 31, 2001.
46. CCIMB Final Interpretation-067, Application Notes Missing in ST, October, 15, 2000.
47. CCIMB Final Interpretation-069, Informal Security Policy Model, March 30, 2001.
48. CCIMB Final Interpretation-074, Duplicate Informative Text for ATE_COV.2-3 and ATE_DPT.1-3, October 15, 2000.
49. CCIMB Final Interpretation-075, Duplicate Informative Text for ATE_FUN.1-4 and ATE_IND.2-1, October 15, 2000.
50. CCIMB Final Interpretation-080, APE_REQ.1-12 Does Not Use "Shall Examine ... to Determine," October 15, 2000.
51. CCIMB Final Interpretation-084, Separate Objectives for TOE and Environment, February 16, 2001.
52. CCIMB Final Interpretation-092, Release of the TOE, July 31, 2001.
53. CCIMB Final Interpretation-094, FLR Guidance Documents Missing, July 31, 2001.
54. CCIMB Final Interpretation-095, ACM_CAP Dependency on ACM_SCP, February 16, 2001.
55. CCIMB Final Interpretation-116, Indistinguishable Work Units for ADO_DEL, July 31, 2001.
56. CCIMB Final Interpretation-120, Indistinguishable Work Units for ADO_DEL, July 31, 2001.
57. CCIMB Final Interpretation-127, Work Unit Not at the Right Place, October 29, 2001.
58. CCIMB Final Interpretation-128, Coverage of the Delivery Procedures, October 29, 2001.
59. CCIMB Final Interpretation-133, Consistency Analysis in AVA_MSU.2, February 16, 2001.

60. EN 45001, General Criteria for the Operation of Testing Laboratories, CEN/CENELEC, 1989.
61. EN 45011, General Criteria for Certification Bodies Operating Product Certification Systems, CEN/CENELEC, 1989.
62. ISO/IEC 13335-1(1996-12), Information Technology — Guidelines for the Management of IT Security — Part 1: Concepts and Models for IT Security.
63. ISO/IEC 13335-2(1997-12), Information Technology — Guidelines for the Management of IT Security — Part 2: Managing and Planning IT Security.
64. ISO/IEC 13335-3(1998-06), Information Technology — Guidelines for the Management of IT Security — Part 3: Techniques for the Management of IT Security.
65. ISO/IEC 13335-4(2000-03), Information Technology — Guidelines for the Management of IT Security — Part 4: Selection of Safeguards.
66. ISO/IEC 13335-5(2001-11), Information Technology — Guidelines for the Management of IT Security — Part 5: Management Guidance on Network Security.
67. ISO/IEC 17025(1999), General Requirements for the Competence of Calibration and Testing Laboratories; superseded ISO/IEC Guide 25(1990).
68. ISO/IEC Guide 65(1996), General Requirements for Bodies Operating Product Certification Systems.
69. ISO/IEC 17799(2000-12), Information Technology — Code of Practice for Information Security Management.
70. ISO/IEC 13233(1995-12), Information Technology — Interpretation of Accreditation Requirements in ISO/IEC Guide 25 — Accreditation of Information Technology and Telecommunications Testing Laboratories for Software and Protocol Testing Services.
71. ISO/IEC 15292(2001-12), Information Technology — Security Techniques — Protection Profile registration procedures.
72. ISO 2382-8(1998-11), ed. 2.0, Information Technology — Vocabulary — Part 8: Security.
73. ISO 9000(2000) Compendium, International Standards for Quality Management.
74. NSTISSP #6: *National Policy on Certification and Accreditation of National Security Telecommunications and Information Systems*, National Security Telecommunications and Information System Security Committee, April 8, 1994.
75. NSTISSP #11: *National Information Assurance Acquisition Policy*, National Security Telecommunications and Information System Security Committee, January 2000.
76. NSTISSAM INFOSEC/2-00, Advisory Memorandum for the Strategy for Using the National Information Assurance Partnership (NIAP®) for the Evaluation of Commercial Off-the-Shelf (COTS) Security Enabled Information Technology Products, February 8, 2000.
77. NSTISSAM COMPUSEC/1-99, Advisory Memorandum on the Transition from the Trusted Computer System Evaluation Criteria (TCSEC) to the International Common Criteria for Information Technology Security Evaluation, March 11, 1999.
78. NSTISSI #1000: *National Information Assurance Certification and Accreditation Process (NIACAP)*, National Security Telecommunications and Information System Security Committee, April 2000.
79. NSTISSI #4009: *National Information System Security (INFOSEC) Glossary*, National Security Telecommunications and Information System Security Committee, January 1999.
80. FIPS PUB 140-2, *Security Requirements for Cryptographic Modules*, National Institute of Standards and Technology, U.S. Department of Commerce, May 25, 2001.
81. FIPS PUB 197, *Advanced Encryption Standard (AES)*, National Institute of Standards and Technology, U.S. Department of Commerce, November 26, 2001.
82. NIST Special Publication 800-21, *Guidelines for Implementing Cryptography in the Federal Government*, National Institute of Standards and Technology, U.S. Department of Commerce, November 1999.

83. NIST Special Publication 800-23, *Guidelines to Federal Organizations on Security Assurance and Acquisition: Use of Tested/Evaluated Products*, National Institute of Standards and Technology, U.S. Department of Commerce, August 2000.
84. Public Law 100-235, Computer Security Act of 1987, January 8, 1988.
85. Public Law 104-106, the Information Technology Management Reform Act of 1996.
86. OMB Circular A-130, *Appendix III: Security of Federal Automated Information Resources*, February 8, 1996.
87. PDD-63, Critical Infrastructure Protection, Presidential Decision Directive, May 13, 1998.
88. *Information Assurance Technical Framework (IATF)*, version 3.0, September 2000.
89. *System Security Engineering Capability Maturity Model (SSE-CMM)*, version 2.0, April 1999.
90. *Systems Security Engineering Capability Maturity Model (SSE-CMM) Appraisal Method*, version 2.0, April 1999.
91. *Common Criteria Toolbox™*, version 6.0f, June 2001.

Publications

92. Abrams, M., Parraga, F., and Veoni, J., Application of the Protection Profile to Define Requirements for a Telecommunications Services Contract, paper presented at the 1st Symposium on Requirements Engineering for Information Security, CERIAS, Purdue University, West Lafayette, IN, March 2001.
93. Belvin, F., Introduction to the CEM, paper presented at the 23rd National Information Systems Security Conference, October 2000.
94. Caplan, K. and Sanders, J., Building an international security standard, *IT Professional*, 1(2), 29–34, 1999.
95. *Common Criteria: An Introduction*, Common Criteria Project Sponsoring Organizations, October 1999.
96. *Common Criteria for Information Technology Security Evaluation: User Guide*, Common Criteria Project Sponsoring Organizations, October 1999.
97. Gertz, B., CIA: Russia, China working on information warfare, *The Washington Times*, June 22, 2001, p. A3.
98. Herrmann, D. and Keith, S., Application of Common Criteria to telecomm services, *Computer Security Journal*, 17(2), 21–28, 2001.
99. Herrmann, D., *A Practical Guide to Security Engineering and Information Assurance*, Auerbach, Boca Raton, FL, 2001.
100. Kekicheff, M., Kashef, F., and Brewer, D., The Open Platform Protection Profile (OP3): Taking the Common Criteria to the Outer Limits, paper presented at the 23rd National Information Systems Security Conference, October 2000.
101. McEvilley, M., Introduction to the Common Criteria, paper presented at the 23rd National Information Systems Security Conference, October 2000.
102. NIAP®, *Organization, Management, and Concept of Operations*, version 2.0, Common Criteria Evaluation and Validation Scheme (CCEVS) for IT Security, Scheme Publication 1, National Information Assurance Partnership, May 1999.
103. NIAP®, *Validation Body Standard Operating Procedures*, draft 1.5, Common Criteria Evaluation and Validation Scheme (CCEVS) for IT Security, Scheme Publication 2, National Information Assurance Partnership, May 2000.
104. NIAP®, *Guidance to Validators of IT Security Evaluations*, Common Criteria Evaluation and Validation Scheme (CCEVS) for IT Security, Scheme Publication 3, National Information Assurance Partnership, February 2002.

105. NIAP®, *Guidance to Common Criteria Testing Laboratories (CCTLs)*, version 1.0, Common Criteria Evaluation and Validation Scheme (CCEVS) for IT Security, Scheme Publication 4, National Information Assurance Partnership, March 20, 2001.
106. NIAP®, *Guidance to Sponsors of IT Security Evaluations*, draft 1.0, Common Criteria Evaluation and Validation Scheme (CCEVS) for IT Security, Scheme Publication 5, National Information Assurance Partnership, August 31, 2000.
107. NIAP®, *Certificate Maintenance Program*, Common Criteria Evaluation and Validation Scheme (CCEVS) for IT Security, Scheme Publication 6, National Information Assurance Partnership, December 2002.
108. *NVLAP® Directory 2002*, NIST Special Publication 810, National Institute of Standards and Technology (note: this document is updated annually).
109. Neumann, P., *Computer Related Risks*, Addison-Wesley, Reading, MA, 1995.
110. *Procedures and General Requirements*, NVLAP® Handbook 150, National Institute of Standards and Technology and U.S. Department of Commerce, July 2001.
111. *Cryptographic Module Testing*, NVLAP® Handbook 150-17, National Institute of Standards and Technology and U.S. Department of Commerce, June 2000.
112. *Information Security Testing: Common Criteria*, version 1.1, NVLAP® (draft) Handbook 150-20, National Institute of Standards and Technology and U.S. Department of Commerce, April 1999.
113. *Written Procedures for NVLAP® Handbook 150-20*, NVLAP® Lab Bulletin LB-5-2001, October 2001.
114. Olthoff, K., Thoughts and Questions on Common Criteria Evaluations, paper presented at the 23rd National Information Systems Security Conference, October 2000.
115. Smith, R., Trends in Government Endorsed Security Product Evaluations, paper presented at the 23rd National Information Systems Security Conference, October 2000.
116. Towns, M., CC Toolbox™, paper presented at the 23rd National Information Systems Security Conference, October 2000.

Online Resources

The following online resources, which were accurate at the time of writing, provide current information about a variety of issues related to the Common Criteria.

117. www.commoncriteria.org; centralized resource for current information about the Common Criteria standards, members, and events.
118. www.iatf.net; Information Assurance Technical Framework standard and forum.
119. www.nstissc.gov/Assets/pdf/nstissi_1000.pdf; National Information Assurance Certification and Accreditation Process (NIACAP).
120. http://niap.nist.gov; National Information Assurance Partnership (NIAP®).
121. http://csrc.nist.gov; National Institute of Standards and Technology (NIST), Computer Security Resource Clearinghouse.
122. http://csrc.nist.gov/crptval; information about NIST cryptographic validation program.
123. www.nist.gov/nvlap; National Voluntary Laboratory Accreditation Program (NVLAP®) publications, including current listing of accredited laboratories.
124. www.radium.ncsc.mil/pep; U.S. Department of Defense Information Security Product Evaluation Programs.
125. http://secinf.net/info/policy/hk_polic.html; computer and information security policy.

126. www.psycom.net/war.1.html; Institute for the Advanced Study of Information Warfare.
127. www.iec.ch.org; International Electrotechnical Commission.
128. www.sse-cmm.org/librarie.htm; latest information about SSE–CMM.
129. www.issea.org; latest information about SSE–CMM.

Annex C: Common Criteria Recognition Agreement (CCRA) Participants

The following organizations, listed in alphabetical order, had signed the Common Criteria Recognition Agreement at the time of writing; it is expected that more countries will sign the agreement in the future. The contact information for each organization is listed. These organizations are the focal point for Common Criteria activities conducted in each of their countries. They are responsible for:*

- Developing, managing, and enforcing their national evaluation scheme in accordance with the CCRA
- Accrediting Common Criteria Evaluation Laboratories in their jurisdiction
- Monitoring and auditing the performance of Common Criteria Evaluation Laboratories in their jurisdiction
- Issuing Common Criteria certificates based on Evaluation Technical Reports
- Recognizing Common Criteria certificates issued by other authorized jurisdictions
- Maintaining the Protection Profile Registry for their jurisdiction
- Maintaining a current Evaluated Products List
- Participating in the Common Criteria Implementation Management Board (CCIMB)

* CCRA participants may be either CC Certificate authorizing participants or CC Certificate consuming participants. CC consuming participants recognize CC Certificates but at present have not implemented a national scheme; this does not preclude them from doing so in the future.

Australia and New Zealand

The Defence Signals Directorate, representing the Federal Government of Australia, and the Government Communications Security Bureau, representing the Government of New Zealand, jointly operate the Australasian Information Security Evaluation Programme (AISEP).

Defence Signals Directorate

AISEP Manager

Information Security Group

Locked Bag 5076

Kingston, ACT 2604

Australia

(telephone) +61.2.6265.0342

(fax) +61.2.6265.0328

(Web site) www.dsd.gov.au/infosec

(e-mail) aisep@dsd.gov.au

Canada

Canadian Common Criteria Evaluation and Certification Scheme

Communications Security Establishment

P.O. Box 9703, Terminal

Ottawa, Ontario

Canada K1G 3Z4

(telephone) +1.613.991.7956

(fax) +1.613.991.7455

(Web site) www.cse-cst.gc.ca/cse/english/cchome.html

(e-mail) ccs-sccc@cse-cst.gc.ca

Finland

Ministry of Finance

P.O. Box 28

00023 Valtioneuvosto

Finland

France

Direction Centrale de la Securite des Systemes d'Information (DCSSI)

Centre de Certification

51 Boulevard de Latour-Maubourg

75700 Paris 07 BP

France

(telephone) +33.1.41463720

(fax) +33.1.41463701

(Web site) www.ssi.gouv.fr

(e-mail) certificaiton,dcssi@sgdn.pm.gouv.fr

Germany

Bundesamt fur Sicherheit in der Informationstechnik

Postfach 20 03 63

53133 Bonn

Germany

(telephone) +49.228.9582.300

(fax) +49.228.9582.427

(Web site) www.bsi.de/cc

(e-mail) cc@bsi.de

Greece

Ministry of Interior

Pan. Kanellopoulou 4

Athens 10177

Greece

Israel

Standards Institution of Israel

42 Lebanon Street

69977 Tel Aviv

Israel

Italy

Autorita Nazionale per la Sicurezza

Via della Pineta Sacchetti N 216

00168 Roma

Italy

The Netherlands

Netherlands National Communications Security Agency

P.O. Box 20061

NL 2500 EB The Hague

The Netherlands

(telephone) +31.70.3485637

(fax) +31.70.3486503

(Web site) www.tno.nl/instit/fel/refs/cc.html

(e-mail) criteria@nbv.cistron.nl

Norway

CHOD Norway/Security Division

HQ Defence Command Norway/Security Division

P.O. Box 14

N-1306 BPD

Norway

Spain

Ministerio de Administraciones Publicas

Maria de Molina, 50

28071 Madrid

Spain

Sweden

Contact information not available at time of publication.

United Kingdom

Certification Body Secretariat

U.K. IT Security Evaluation and Certification Scheme

P.O. Box 152

Cheltenham GL52 5UF

United Kingdom

(telephone) +44.1242.238739

(fax) +44.1242.235233

(Web site) www.cesg.gov.uk/assurance/iacx/itsec/index.htm

(e-mail) info@itsec.gov.uk

United States

National Information Assurance Partnership (NIAP®)

100 Bureau Drive (Mailstop 8930)

Gaitherburg, Maryland 20899–8930

(telephone) +1.301.975.2934

(fax) +1.301.948.0279

(Web site) http://niap.nist.gov/cc-scheme

(e-mail) scheme-comments@nist.gov

Annex D: Accredited Common Criteria Testing Labs

The following organizations, listed in alphabetical order, were recognized as accredited Common Criteria Testing Laboratories by their National Evaluation Authorities and Common Criteria Recognition Agreement (CCRA) participants at the time of writing.

Australia and New Zealand

Computer Sciences Corporation (CSC) Australia

Ground Floor, 15 National Circuit

Barton, Canberra 2600

Australia

(telephone) +61.2.6270.8300

(fax) +61.2.6270.8492

(Web site) www.csc.com.au

(e-mail) aisef@csc.com.au

***Status:* AISEP accreditation for Common Criteria and ITSEC evaluations, NATA approved lab, code 13259**

CMG Admiral

Suite 2, 26–28 Napier Close

Deakin, ACT 2600

Australia

(telephone) +61.2.6211.2000

(fax) +61.2.6260.4255

(Web site) www.admiral.com.au

(e-mail) aisef@au.com

***Status:* AISEP accreditation for Common Criteria and ITSEC evaluation, NATA approved lab**

Canada

CGI

1400–275 Slater Street, 14th Floor

Ottawa, Ontario

Canada K1P 5H9

(telephone) +1.613.234.2155

(fax) +1.613.234.6934

(Web site) www.cgo.ca

(e-mail) andrew.pridham@cgi.ca

***Status:* Licensed Canadian Common Criteria Evaluation Facility**

DOMUS IT Security Laboratory, IBM Canada Ltd.

2300 St. Laurent Blvd.

Ottawa, Ontario

Canada K1G 5L2

(telephone) +1.613.247.5505

(fax) +1.613.739.4936

(Web site) www.domusitsl.com

(e-mail) Lauriem@ca.ibm.com

***Status:* Licensed Canadian Common Criteria Evaluation Facility**

EWA-Canada

275 Slater Street

Ottawa, Ontario

Canada K1P 5H9

(telephone) +1.613.230.6067 ext. 227

(fax) +1.613.230.4933

(Web site) www.ewa-canada.com

(e-mail) pzatychec@ewa-canada.com

Status: **Licensed Canadian Common Criteria Evaluation Facility**

France

AQL-Groupe SILICOMP

Rue de la Chataigneraic BP 127

35513 Cesson Sevigne

Cedex, France

(telephone) +33.(0).2.99.12.50.00

(fax) +33.(0).2.99.63.70.40

(web site) www.aql.fr/AQL SSI CESTI.htm

CEACI (THALES CNES)

18, avenue Edoard Belin

Cedex 4, France

(telephone) +33.(0).5.61.27.40.29

(fax) +33.(0).5.61.27.47.32

SERMA TECHNOLOGIES

30, avenue Gustave Eifel

33608 Pessac

Cedex , France

(telephone) +33.(0).5.57.26.08.64

(fax) +33.(0).5.57.26.08.98

CEA-Leti

17, rue des martyrs

38054 Grenoble

Cedex 9, France

(telephone) +33.(0).4.38.78.40.87

(fax) +33.(0).4.38.78.51.59

Germany

debis Systemhaus Information Security Services GmbH

Rabinstrasse 8

D-53111 Bonn

Germany

(telephone) +49.228.9841.115

(fax) +49.228.9841.60

(Web site) www.itsec-debis.de

(e-mail) wolfgang.killmann@t-systems.de

Status: **CLEF for Common Criteria and ITSEC evaluations, per German Information Security Agency (BSI)**

SEELAB

TUV Nord e.V.

Gr. Bahnstrasse 31

22525 Hamburg

Germany

(telephone) +49.40.8557.2288

(fax) +49.40.8557.2429

(Web site) www.tuev-nord.de/leistungliteng/SEELAB_ST.HTM

(e-mail) seelab@tuev-nord.de

Status: **CLEF for Common Criteria and ITSEC evaluations, per German Information Security Agency (BSI)**

TUV Informationstechnik GmbH

Am Technologiepark 1

45307 Essen

Germany

(telephone) +49.201.8999.624

(fax) +49.201.8999.666

(Web site) www.tuvit.de

(e-mail) W.Peter@tuvit.de

Status: **CLEF for Common Criteria and ITSEC evaluations, per German Information Security Agency (BSI)**

United Kingdom

Admiral Management Services Ltd.

Kings Court

91–93 High Street

Camberley

Surrey GU15 3RN

United Kingdom

(telephone) +44.1276.686678

(fax) +44.1276.691028

(e-mail) worsw_r@admiral.co.uk

***Status:* CLEF for Common Criteria and ITSEC evaluations**

EDS Ltd.

Wavendon Tower

Wavendon

Milton Keynes

Buckinghamshire MK17 8LX

United Kingdom

(telephone) +44.1908.284234

(fax) +44.1908.284393

(e-mail) trevor.hutton@edl.uk.eds.com

***Status:* CLEF for Common Criteria and ITSEC evaluations**

IBM

Meudon House, Meudon Avenue

Farnborough

Hannts GU14 7NB

United Kingdom

(telephone) +44.1252.558081

(fax) +44.1252.558001

(Web site) www.ibm.com/security/services/consult-

(e-mail) bob_finlay@uk.ibm.com

***Status:* Licensed CLEF, accredited by UKAS**

Security Practice & CLEF, Logica U.K., Ltd.

Chaucer House, The Office Park

Springfield Drive

Leatherhead

Surrey KT22 7LP

United Kingdom

(telephone) +44.1372.369830

(fax) +44.1372.369834

(Web site) www.logica.com/security

(e-mail) clef@logica.com

***status:* CLEF for Common Criteria and ITSEC evaluations, accredited by UKAS, lab code 1097**

SYNTEGRA

Guidion House

Ancells Business Park

Fleet

Hampshire GU13 8UZ

United Kingdom

(telephone) +44.1252.778837

(fax) +44.1252.811635

(Web site) www.syntefra.com

(e-mail) clef@syntegra.bt.co.uk

***Status:* CLEF for Common Criteria and ITSEC evaluations**

United States

Arca Systems, Inc.

10220 Old Columbia Road

Suite G–H

Columbia, Maryland 21046–2366

(telephone) +1.410.309.1780

(fax) +1.410.309.1781

(Web site) www.arca.com

(e-mail) diann.carpene

Status: **NVLAP® accreditation for Common Criteria testing, lab code 200429–0; NIAP® CCEVS approved Common Criteria testing lab**

Booz Allen Hamilton

900 Elkridge Landing Road, Suite 100

Linthicum, Maryland 21090

(telephone) +1.410.684.6602

(fax) +1.410.309.6475

(e-mail) rome-steven@bah.com

Status: **NVLAP® accreditation for Common Criteria testing, lab code 200423–0; NIAP® CCEVS approved Common Criteria testing lab**

COACT, Inc.

9140 Guilford Road, Suite L

Columbia, Maryland 21046

(telephone) +1.301.498.0150

(fax) +1.301.498.0855

(Web site) www.coact.com

(e-mail) ejg@coact.com

Status: **NVLAP® accreditation for Common Criteria testing and for Cryptographic Module Testing (FIPS 140-2), lab code 200416-0; NIAP® CCEVS approved Common Criteria testing lab**

Computer Sciences Corporation

132 National Business Parkway, 4th Floor

Annapolis Junction, Maryland 20701

(telephone) +1.240.456.6227

(Web site) www.csc.com/ttap

(e-mail) jfink5@csc.com

Status: **NVLAP® accreditation for Common Criteria testing, lab code 200426–0; NIAP® CCEVS approved Common Criteria testing lab**

CygnaCom

7927 Jones Branch Drive

Suite 100 West

McLean, Virginia 22102–3305

(telephone) +1.703.848.0883

(fax) +1.703.848.0960

(Web site) www.entrust.com/entrustcygnacom/labs/sel.htm

(e-mail) jthompson@cygna.com.com

***Status:* NVLAP® accreditation for Common Criteria testing, lab code 200002-0; NIAP® CCEVS approved Common Criteria testing lab**

Science Applications International Corporation (SAIC)

7125 Columbia Gateway Drive

Suite 300

Columbia, Maryland 21046–2554

(telephone) +1.410.953.6819

(fax) +1.410.953.6930

(Web site) www.cist.saic.com/index.html

(e-mail) robert.l.williamson.jr@cpmx.saic.com

***Status:* NVLAP® accreditation for Common Criteria testing, lab code 200427-0; NIAP® CCEVS approved Common Criteria testing lab**

Annex E: Accredited Cryptographic Module Testing Laboratories

The following organizations, listed in alphabetical order, were recognized as Accredited Cryptographic Module Testing laboratories in accordance with the Cryptographic Module Validation Program (CMVP) at the time of writing. For a current list, consult the NVLAP® Web site. (Note: in general, laboratories must be re-accredited every two years.)

Canada

DOMUS IT Security Laboratory

2220 Walkley Road

Ottawa, Ontario

Canada K1G 5L2

(telephone) +1.613.247.5505

(fax) +1.613.230.3274

(Web site) www.domusitsl.com

(e-mail) lauriem@ca.ibm.com

***Status:* accreditation valid through December 31, 2001, for:**

- 17/C01 — NIST-CSTT:140-1; National Institute of Standards and Technology — Cryptographic Support Test Tool for FIPS PUB 140-1, Security Requirements for Cryptographic Modules
- 17/C01a — Test Method Group 1: All test methods derived from FIPS PUB 140-1 and specified in the CSTT, except those listed in Group 2 and Group 3
- 17/C01b — Test Method Group 2: Test methods for Physical Security, Level 4, derived from FIPS PUB 140-1 and specified in the CSTT
- 17/C01c — Test Method Group 3: Test methods for Software Security, Level 4, derived from FIPS PUB 140-1 and specified in the CSTT
- 17/C02 — FIPS-Approved Cryptographic Algorithms as required in FIPS PUB 140-1

United States

Atlan Laboratories

1340 Old Chain Bridge Road, Suite 401

McLean, Virginia 22101

(telephone) +1.703.748.4551

(fax) +1.703.748.4552

(Web site) www.atlanlabs.com

(e-mail) emorris@atlanlabs.com

***Status:* accreditation valid through December 31, 2001, for:**

- 17/C01 — NIST-CSTT:140-1; National Institute of Standards and Technology — Cryptographic Support Test Tool for FIPS PUB 140-1, Security Requirements for Cryptographic Modules
- 17/C01a — Test Method Group 1: All test methods derived from FIPS PUB 140-1 and specified in the CSTT, except those listed in Group 2 and Group 3
- 17/C01b — Test Method Group 2: Test methods for Physical Security, Level 4 derived from FIPS PUB 140-1 and specified in the CSTT
- 17/C01c — Test Method Group 3: Test methods for Software Security, Level 4 derived from FIPS PUB 140-1 and specified in the CSTT
- 17/C02 — FIPS-Approved Cryptographic Algorithms as required in FIPS PUB 140-1

COACT Inc. CAFÉ Laboratory

9140 Guilford Road, Suite L

Columbia, Maryland 21046

(telephone) +1.301.498.0150

(fax) +1.301.498.0855

(Web site) www.coact.com

(e-mail) jom@coact.com

Status: **accreditation valid through December 31, 2001, for:**

- 17/C01 — NIST-CSTT:140-1; National Institute of Standards and Technology — Cryptographic Support Test Tool for FIPS PUB 140-1, Security Requirements for Cryptographic Modules
- 17/C01a — Test Method Group 1: All test methods derived from FIPS PUB 140-1 and specified in the CSTT, except those listed in Group 2 and Group 3
- 17/C02 — FIPS-Approved Cryptographic Algorithms as required in FIPS PUB 140-1

Cygnacom Solutions, Inc., An Entrust Company

7929 Jones Branch Drive, Suite 100 West

McLean, Virginia 22102-3305

(telephone) +1.703.270.3520

(fax) +1.703.848.0960

(Web site) www.cygnacom.com/labs/sel.htm

(e-mail) chokhani@cygnacom.com

Status: **accreditation valid through September 30, 2002, for:**

- 17/C01 — NIST-CSTT:140-1; National Institute of Standards and Technology — Cryptographic Support Test Tool for FIPS PUB 140-1, Security Requirements for Cryptographic Modules
- 17/C01a — Test Method Group 1: All test methods derived from FIPS PUB 140-1 and specified in the CSTT, except those listed in Group 2 and Group 3
- 17/C01b — Test Method Group 2: Test methods for Physical Security, Level 4, derived from FIPS PUB 140-1 and specified in the CSTT
- 17.C01c — Test Method Group 3: Test methods for Software Security, Level 4, derived from FIPS PUB 140-1 and specified in the CSTT
- 17/C02 — FIPS-Approved Cryptographic Algorithms as required in FIPS PUB 140-1

InfoGard Laboratories, Inc.

641 Higuera Street, Second Floor

San Luis Obispo, California 93401

(telephone) +1.805.783.0810

(fax) +1.805.783.0889

(Web site) www.infogard.com

(e-mail) mbrinton@infogard.com

***Status:* accreditation valid through June 30, 2002, for:**

- 17/C01 — NIST-CSTT:140-1; National Institute of Standards and Technology — Cryptographic Support Test Tool for FIPS PUB 140-1, Security Requirements for Cryptographic Modules
- 17/C01a — Test Method Group 1: All test methods derived from FIPS PUB 140-1 and specified in the CSTT, except those listed in Group 2 and Group 3
- 17/C01b — Test Method Group 2: Test methods for Physical Security, Level 4, derived from FIPS PUB 140-1 and specified in the CSTT
- 17/C01c — Test Method Group 3: Test methods for Software Security, Level 4, derived from FIPS PUB 140-1 and specified in the CSTT
- 17/C02 — FIPS-Approved Cryptographic Algorithms, as required in FIPS PUB 140-1

Annex F: Glossary of Classes and Families

This glossary provides the long names for functional and assurance class and family three-character mnemonics. A few family mnemonics have multiple meanings; they are used by multiple classes. In this situation, all long names are provided in alphabetical order by class.

ACC	User data protection access control policy
ACF	User data protection access control functions
ACM	Configuration management assurance class
ADM	Guidance documents, administrator guidance
ADO	Delivery and operation assurance class
ADV	Development assurance class
AFL	Authentication failures
AGD	Guidance documents assurance class
ALC	Lifecycle support assurance class
AMT	Protection of the TSF, underlying abstract machine test
ANO	Privacy, anonymity
APE	Protection profile evaluation assurance class
ARP	Security audit automatic response
ASE	Security Target evaluation assurance class
ATD	Identification and authentication user attribute definition
ATE	Tests assurance class
AUT	CM automation
AVA	Vulnerability assessment assurance class
CAP	CM capabilities
CCA	Vulnerability analysis, covert channel analysis
CKM	Cryptographic key management
COP	Cryptographic operation

COV Tests, coverage
DAU User data protection data authentication
DEL Delivery and operation, delivery
DES (1) Protection Profile evaluation, TOE description; (2) Security Target Evaluation, TOE description
DPT Tests, depth
DVS Lifecycle support, development security
ENV (1) Protection Profile evaluation, security environment; (2) Security Target evaluation, security environment
ETC User data protection export to outside TSF control
FAU Security audit functional class
FCO Communication functional class
FCS Cryptographic support functional class
FDP User data protection functional class
FIA Identification and authentication functional class
FLR Lifecycle support, flaw remediation
FLS Protection of the TSF, failure secure
FLT Resource utilization, fault tolerance
FMT Security management functional class
FPR Privacy functional class
FPT Protection of the TSF functional class
FRU Resource utilization functional class
FSP Development, functional specification
FTA TOE access functional class
FTP Trusted path/channels functional class
FUN Tests, functional tests
GEN Security audit generation
HLD Development, high-level design
IFC User data protection information flow control policy
IFF User data protection information flow control functions
IGS Delivery and operation, installation, generation, and start-up
IMP Development, implementation representation
IND Tests, independent testing
INT (1) Protection Profile evaluation, PP introduction; (2) Security Target evaluation, ST introduction; (3) development, TSF internals
ITA Protection of the TSF, availability of exported TSF data
ITC (1) User data protection, import from outside TSF control; (2) protection of the TSF, confidentiality of exported TSF data; (3) trusted path/channels, inter-TSF trusted channel
ITT (1) User data protection, internal TOE transfer; (2) protection of the TSF, internal TOE TSF data transfer
LCD Lifecycle support, lifecycle definition
LLD Development, low-level design
LSA TOE access, limitation on scope of selectable attributes

MCS	TOE access, limitation on multiple concurrent sessions
MOF	Security management, management of functions in TSF
MSA	Security management, management of security attributes
MSU	Vulnerability assessment, misuse
MTD	Security management, management of TSF data
NRO	Communication non-repudiation of origin
NRR	Communication non-repudiation of receipt
OBJ	(1) Protection Profile evaluation, security objectives; (2) Security Target evaluation, security objectives
PHP	Protection of the TSF, TSF physical protection
PPC	Security Target evaluation, PP claims
PRS	Resource utilization, priority of service
PSE	Privacy, pseudonymity
RCR	Development, representation correspondence
RCV	Protection of the TSF, trusted recovery
REQ	(1) Protection Profile evaluation, IT security requirements; (2) Security Target evaluation, IT security requirements
REV	Security management, revocation
RIP	User data protection residual information protection
RPL	. Protection of the TSF; replay detection
ROL	User data protection rollback
RSA	Resource utilization, resource allocation
RVM	Protection of the TSF, reference mediation
SAA	Security audit analysis
SAE	Security management, security attribute expiration
SAR	Security audit review
SCP	CM scope
SDI	User data protection, stored data integrity
SEL	Security audit event selection
SEP	Protection of the TSF, domain separation
SMR	Security management, security management roles
SOF	Vulnerability assessment, strength of TOE security functions
SOS	Identification and authentication specification of secrets
SPM	Development, security policy modeling
SRE	(1) Protection Profile evaluation, explicitly stated IT security requirements; (2) Security Target evaluation, explicitly stated IT security requirements
SSP	Protection of the TSF, state synchrony protocol
SSL	TOE access, session locking
STG	Security audit event storage
STM	Protection of the TSF, time stamps
TAB	TOE access, TOE access banners
TAH	TOE access, TOE access history
TAT	Lifecycle support, tools and techniques

TDC	Protection of the TSF, inter-TSF TSF data consistency
TRC	Protection of the TSF, internal TOE TSF data replication consistency
TRP	Trusted path/channels, trusted path
TSE	TOE access, TOE session establishment
TSS	Security Target evaluation, TOE summary specification
TST	Protection of the TSF, TSF self test
UAU	User authentication
UID	User identification
UNL	Privacy, unlinkability
UNO	Privacy, unobservability
USB	Identification and authentication user–subject binding
USR	Guidance documents, user guidance
VLA	Vulnerability assessment, vulnerability analysis

Index

D

E

F

G

I

L

M

N

O

P

R

S

T

U

V

W